PROJECT M²: MENTORING YOUNG MATHEMATICIANS

Exploring Shape Games

Geometry with Imi and Zani

M. Katherine Gavin, *University of Connecticut*
Tutita M. Casa, *University of Connecticut*
Suzanne H. Chapin, *Boston University*
Linda Jensen Sheffield, *Northern Kentucky University*

Student Mathematician's Journal

Connections to the Common Core State Standards

This material is based upon work supported by the National Science Foundation under Grant No. 0733189 and developed at the University of Connecticut. Any opinions, findings, and conclusions or recommendations expressed in this material are those of the author(s) and do not necessarily reflect the views of the National Science Foundation.

Kendall Hunt
publishing company

www.kendallhunt.com
Send all inquiries to:
4050 Westmark Drive
Dubuque, IA 52004-1840
1-800-542-6657

ISBN 978-1-5249-3127-8

Published in the United States of America

17 16 15 14 13

Production Date: 2016
Printed by: LSI
United States of America
Batch number: 441196

Table of Contents

Speakers' Scroll 1
Listeners' Scroll 2
Writers' Scroll 3

Unit Introduction Lesson
What's the Same About These Shapes? 5
Think Deeply 13

Chapter 1: Together and Apart
Lesson 1: Rectangles and Triangles Together
My Shape 15
Grupo Game Rules 18
Think Deeply 22

Lesson 2: Moving or Removing Shapes
Think Deeply 26

Chapter 2: Playing with Attributes
Lesson 1: That's My Property
Angle Sort 29
One-Loop Puzzle 31
Imi's Two-Loop Puzzles 33
Think Deeply 35

Lesson 2: Shapes of All Sorts
Angle-Sorting Tree 37
Size- and Color-Sorting Tree 39
Angle- and Size-Sorting Tree 41
What's Missing? — A 43
What's Missing? — B 45
What's Missing? — 1 47

Table of Contents

What's Missing? — 2 49
What's Missing? — 3 51
What's Missing? — 4 53
What's Missing? — 5 55
What's Missing? — 6 57
Yes or No 59
Think Deeply 62

Chapter 3: Congruence and Symmetry

Lesson 1: Match It Up!

Same Size and Shape? 65
Think Deeply 67

Lesson 2: Symmetrical Shapes

Where's Imi? 69
Mirror Explorations 71
Do You See What You Started With? 75
Symmetrical Shapes? 77
Pattern Block Pictures 79
Pattern Block Symmetry Answer Sheet 83
Think Deeply 87

Resources

Student Mathematician's Glossary 89

As **Speakers** we will:

1. Talk loudly.

2. Turn to the class.

3. Share and explain our ideas.

4. Agree and disagree with ideas, not with each other.

As **Listeners** we will:

1. Ask speakers to speak up.

2. Show speakers we are listening.

3. Ask questions to understand an idea.

As Writers we will:

about
the question.

about
the answer.

1. all ideas

2. the answer

3. "why"

Student Mathematician: ______________________ Date: __________

What's the Same About These Shapes?

All triangles have ______________________

____

____

____

Student Mathematician: ______________________ Date: __________

What's the Same About These Shapes?

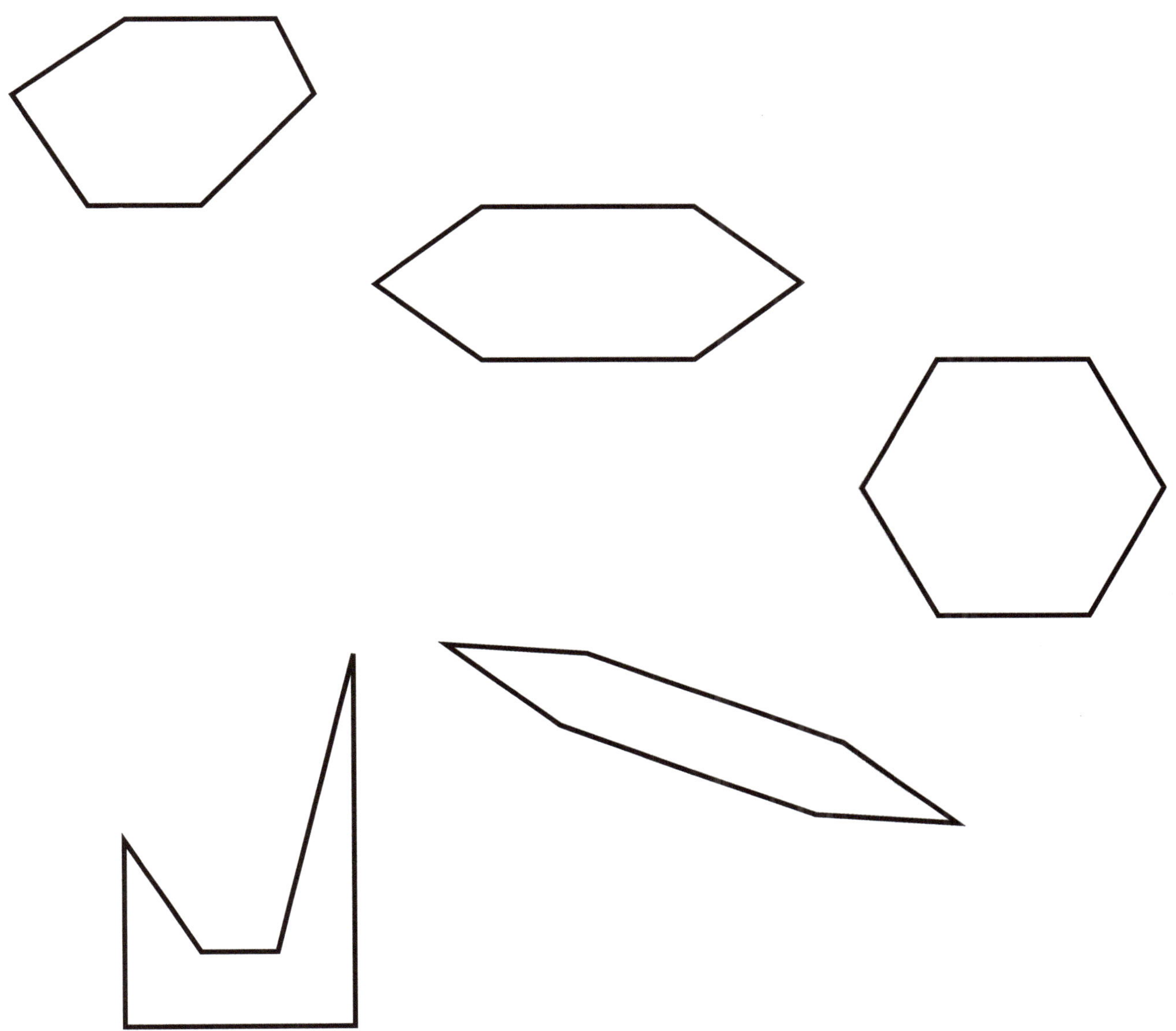

All hexagons have ______________________

____

____

____

Student Mathematician: ______________________ Date: __________

What's the Same About These Shapes?

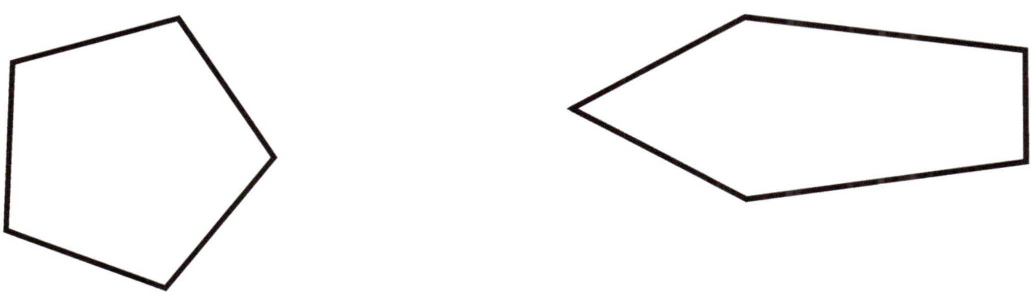

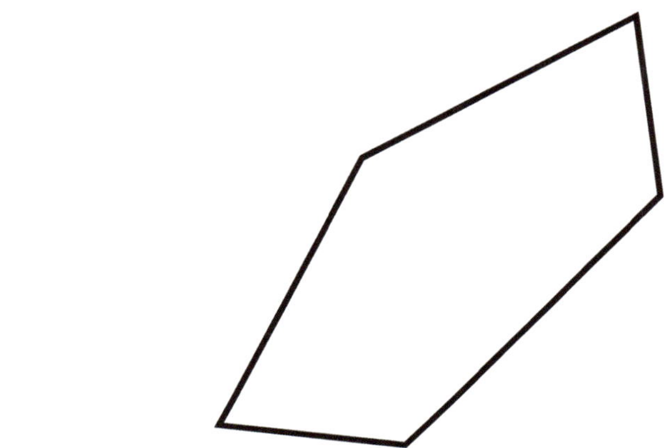

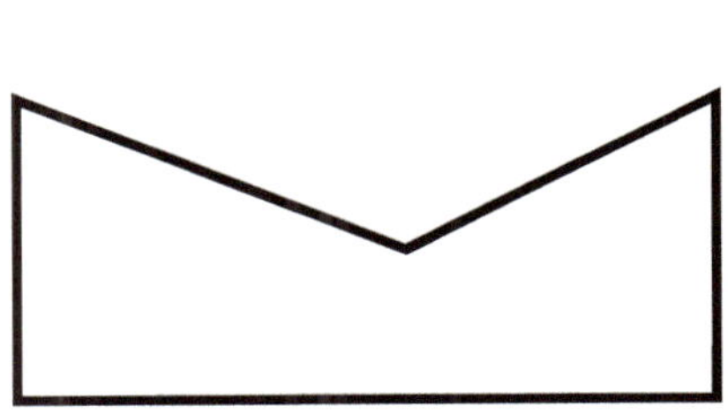

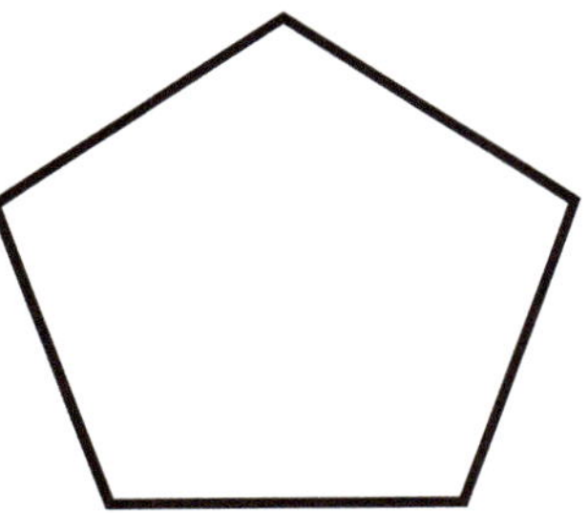

All pentagons have ______________________

____

____

____

Student Mathematician: ________________________ Date: __________

What's the Same About These Shapes?

All quadrilaterals have ______________________________

____

____

____

Student Mathematician: ______________________ Date: __________

THINK DEEPLY

I want to win my Lingo game! I rolled a 4. Write an **X** to show me where to put my next chip. Why should I put it there?

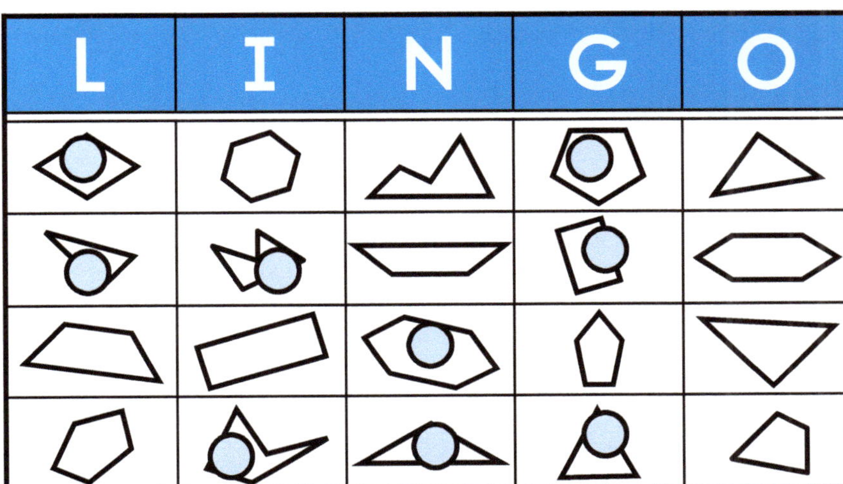

Dear Zani,

Put your chip on the ______________________

____

because ______________________

____

______________________. Now you can win!

Your math friend,

Student Mathematician: ______________________ Date: __________

My Shape

Paste your new shape below. Trace each side. Place one dot on each vertex.

My shape has ______________________ sides.

My shape has ______________________ vertices.

My shape is made of ______________________.

My shape is a ______________________.

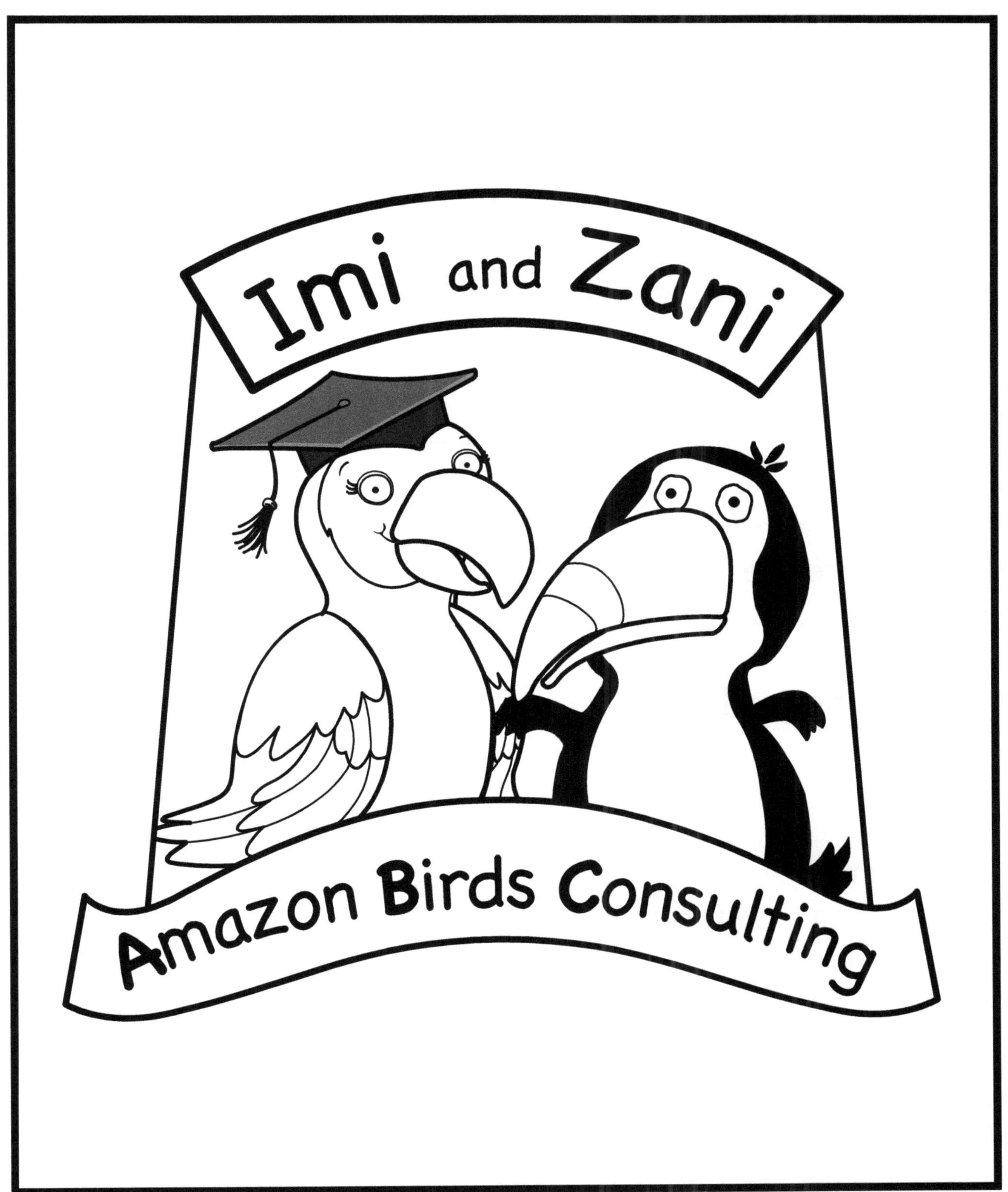
Imi and Zani
Amazon Birds Consulting

Grupo Game Rules

1. The dealer and two teams of partners — 5 friends in all — play this game.

2. The dealer puts 9 cards facing **up** on the grid.

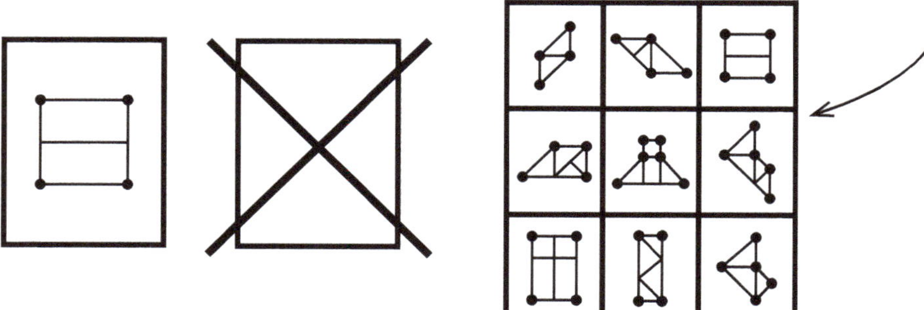

3. Teams take turns making pairs of cards that have 4 things that are the same.

- Same number of **sides**

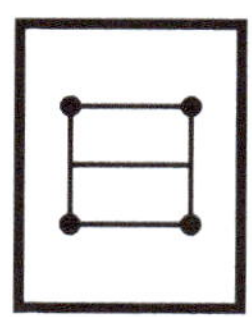 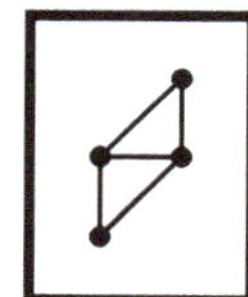

- Same **number of inside shapes**

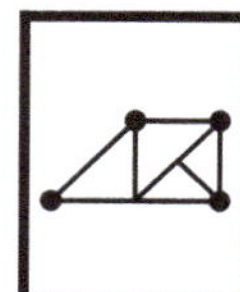 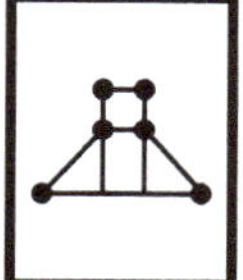

- Same number of **vertices**

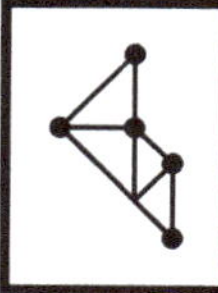 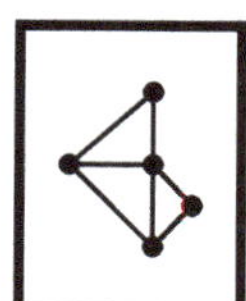

- Same **kind of inside shapes**

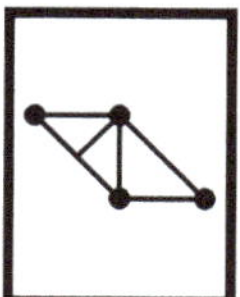 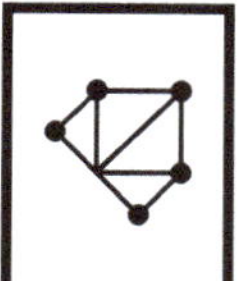

- Same shape **name**

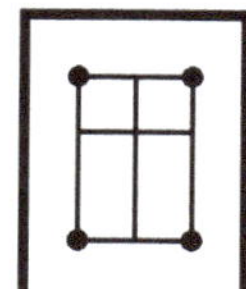 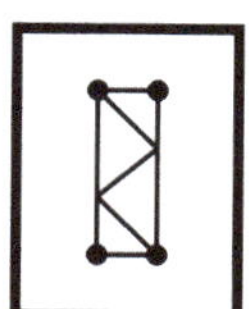

4. Tell the other team YES or NO if this is a pair or not. If YES, pick up the two cards.

5. The dealer adds two new cards to the grid.

6. It is now the other team's turn. Pass if you cannot make a pair. If no one can make a pair, the dealer collects all 9 cards and puts down all new cards.

7. Play until there are no more cards.

8. The team with the most cards wins.

Grupo Checklist

- ☐ Same number of **sides**
- ☐ Same number of **vertices**
- ☐ Same shape **name**
- ☐ Same **number of inside shapes**
- ☐ Same **kind of inside shapes**

Imi and Zani
Amazon Birds Consulting

Student Mathematician: ______________________________ Date: ______________

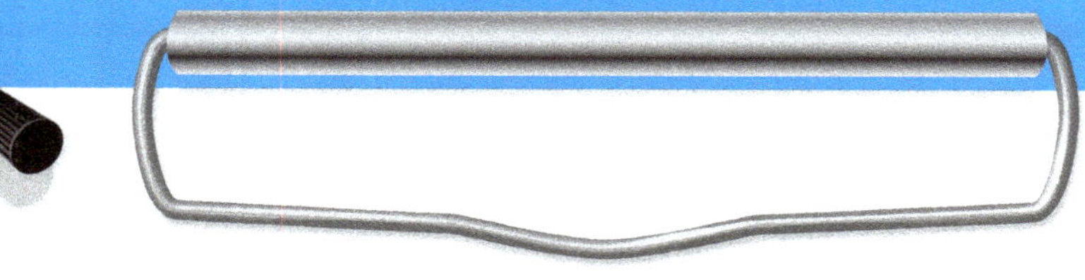

Circle one of the cards.

A

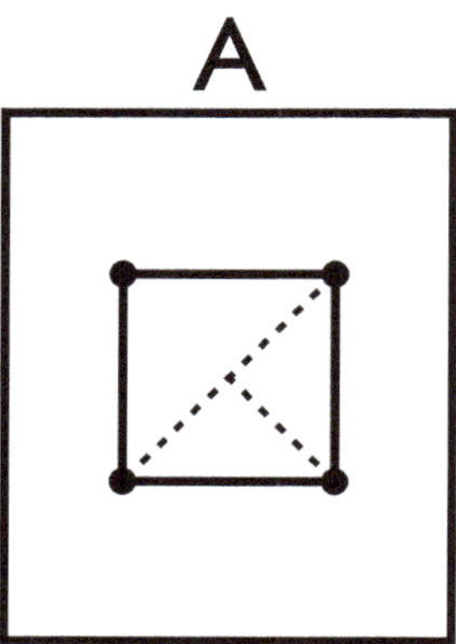

B

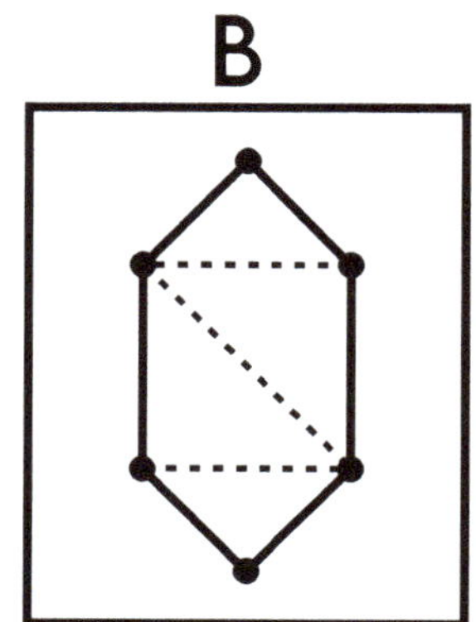

Make a shape that is like the card you circled in FOUR ways.

Glue it below:

Tell us your ideas!

Student Mathematician: ______________________ Date: __________

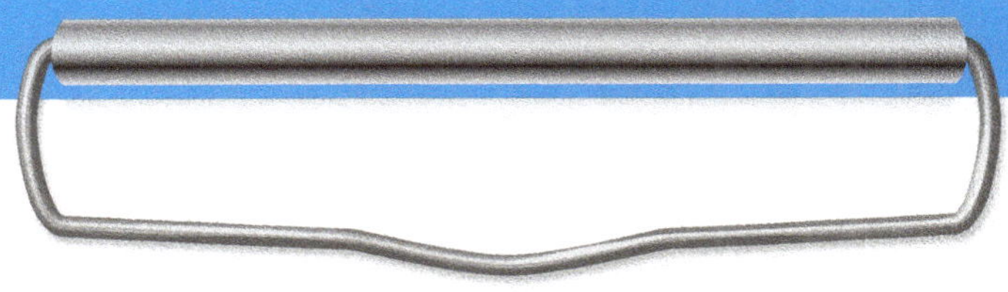

How is your shape <u>like</u> the shape on the card?

1. ______________________

2. ______________________

3. ______________________

4. ______________________

How is your shape <u>different</u> from the shape on the card?

____

____

____

____

____

____

Imi and Zani
Amazon Birds Consulting

Student Mathematician: ______________________ Date: __________

THINK DEEPLY

It was a ______________. Now it is a ______________.

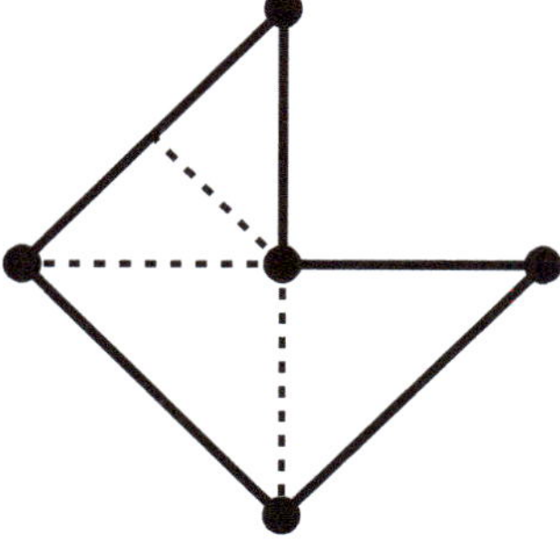

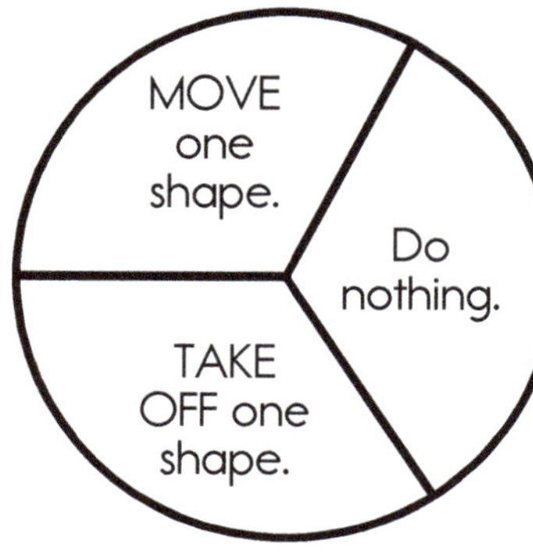

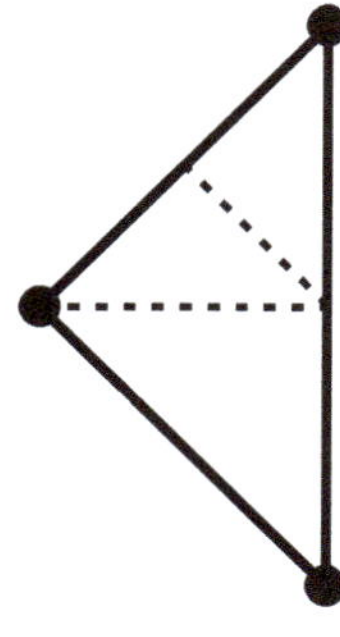

Circle what the spinner said.

It was a ______________. Now it is a ______________.

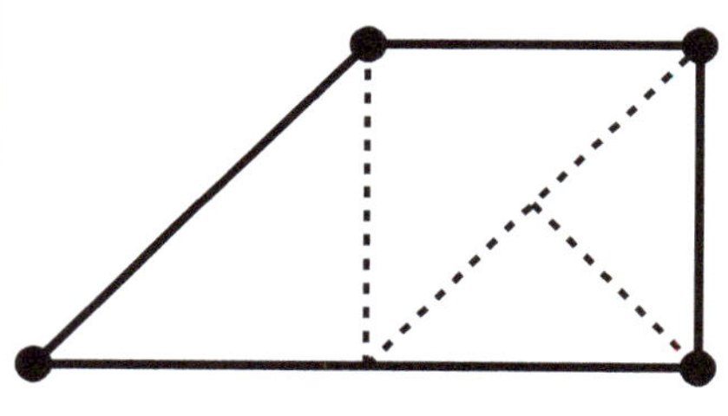

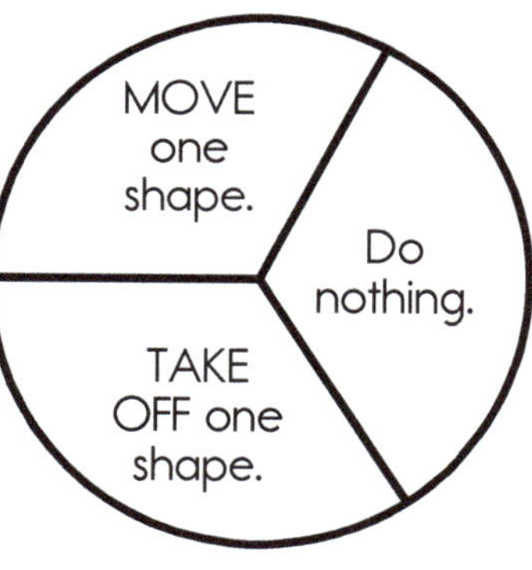

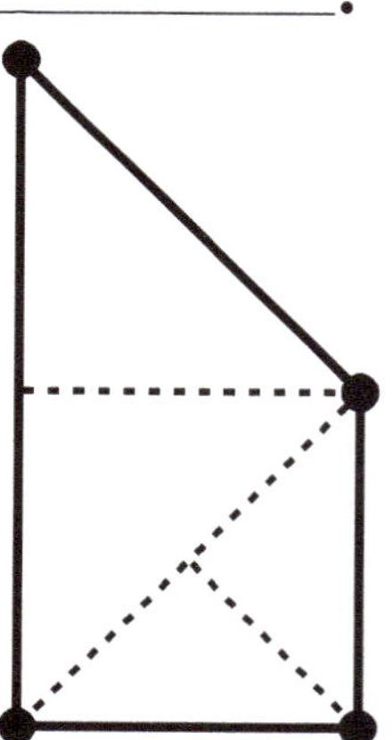

Circle what the spinner said.

Tell us your ideas!

Student Mathematician: ______________________ Date: __________

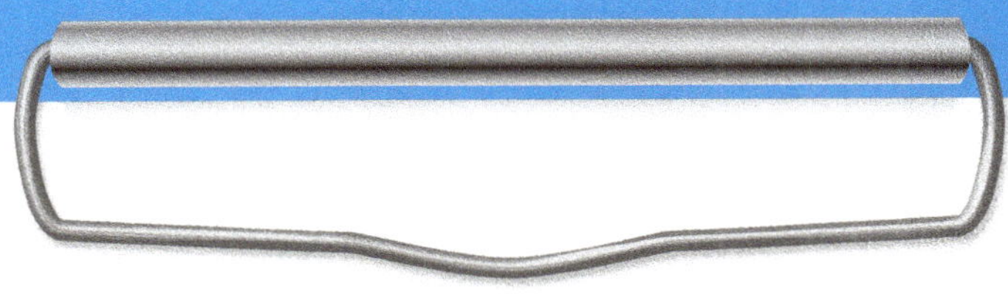

Tell us what strategies you used to play What Happened?

Student Mathematician: ______________________ Date: __________

Angle Sort

Trace a shape that fits each description. Mark right angles green, obtuse angles blue, and acute angles red.

1. All right angles

2. All obtuse angles

3. Some acute angles

Student Mathematician: ______________________ Date: __________

One-Loop Puzzle

Make us a puzzle!

1. Pick one attribute.
2. Draw three shapes inside the loop.
3. Draw three shapes outside the loop.
4. Keep your card and trade papers with a partner.

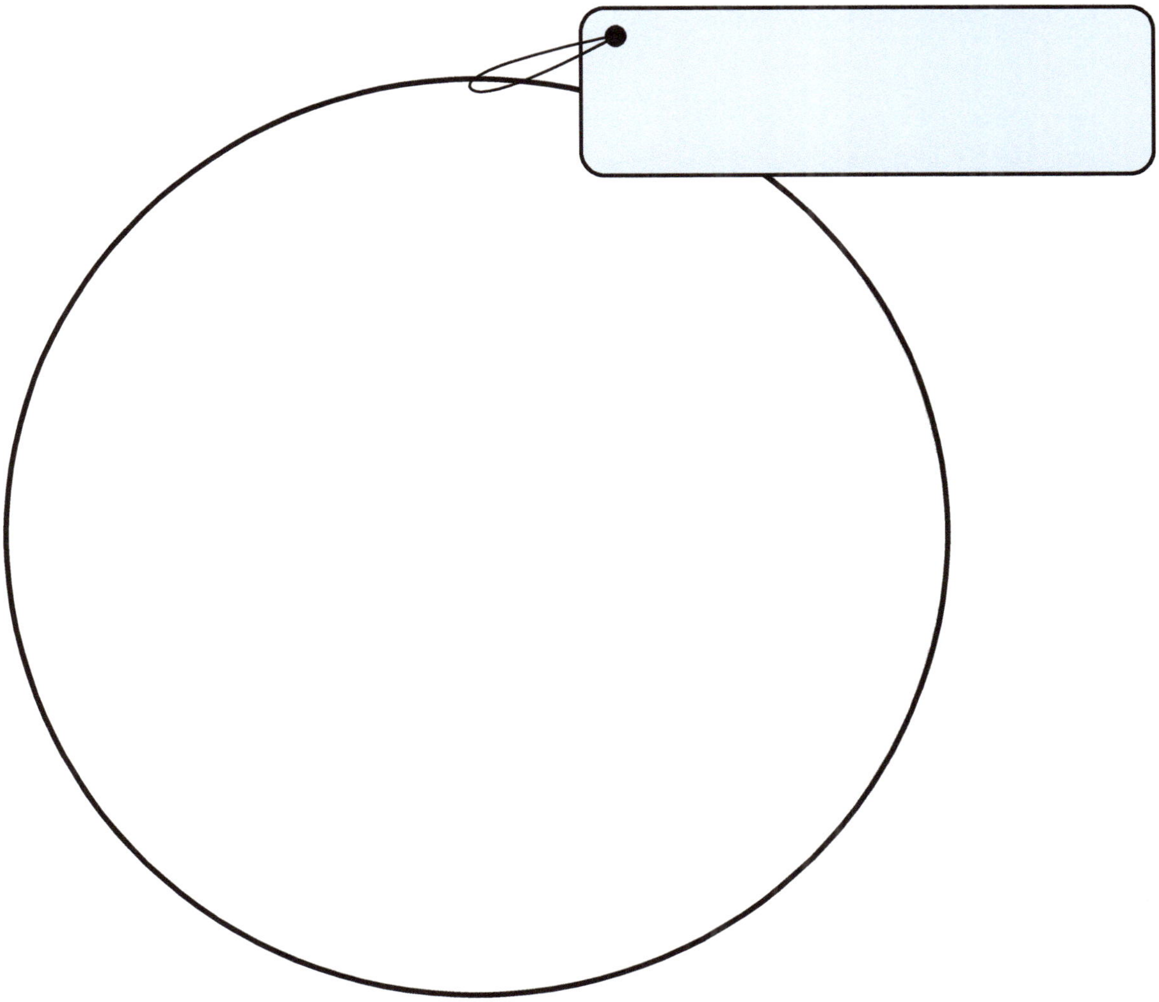

Partner

5. Guess the attribute! Write it in the label.

Student Mathematician: ______________________ Date: ____________

Imi's Two-Loop Puzzles

1. Label both of my loops.

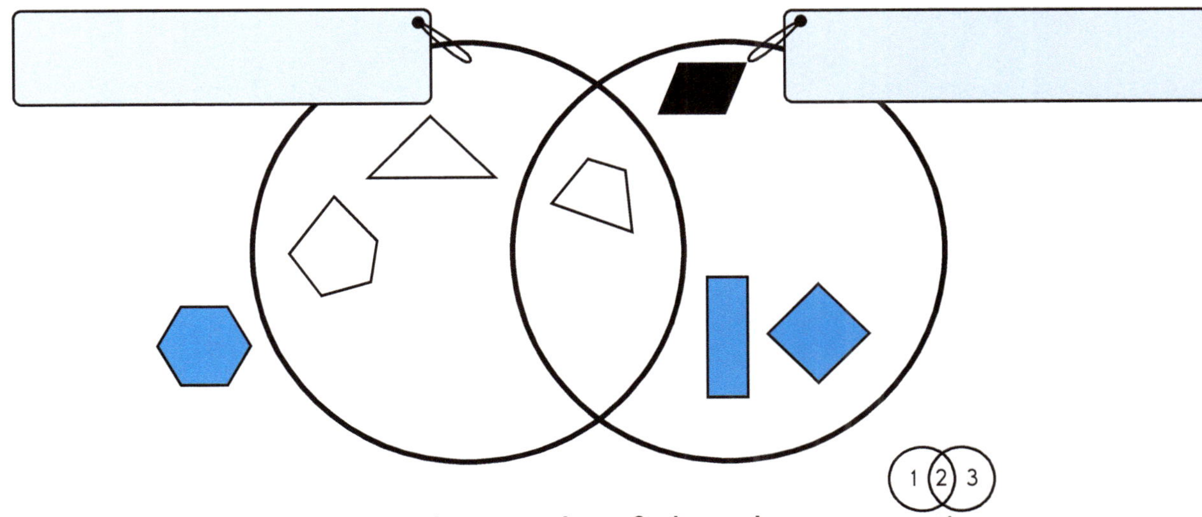

2. Draw a shape inside each of the three sections.
3. Draw a shape outside the loops.

---

1. Label both of my loops.

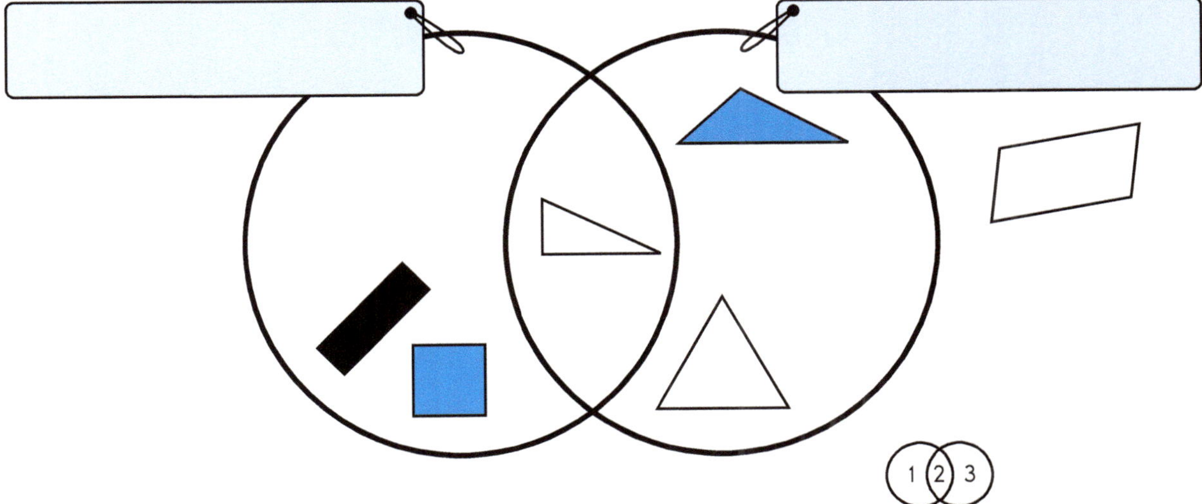

2. Draw a shape inside each of the three sections.
3. Draw a shape outside the loops.

Student Mathematician: ______________________ Date: __________

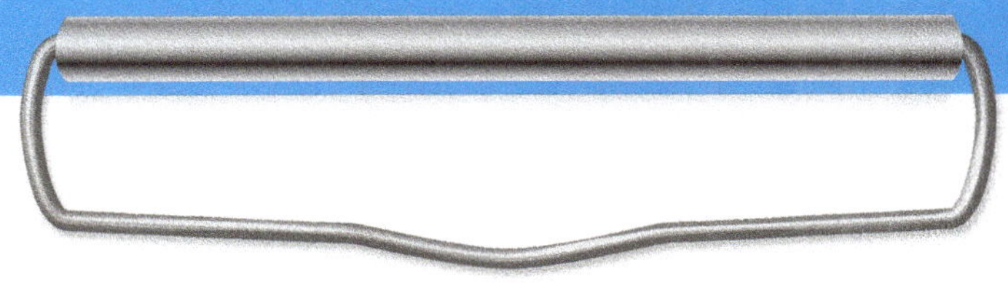

I am puzzled about these two loops!

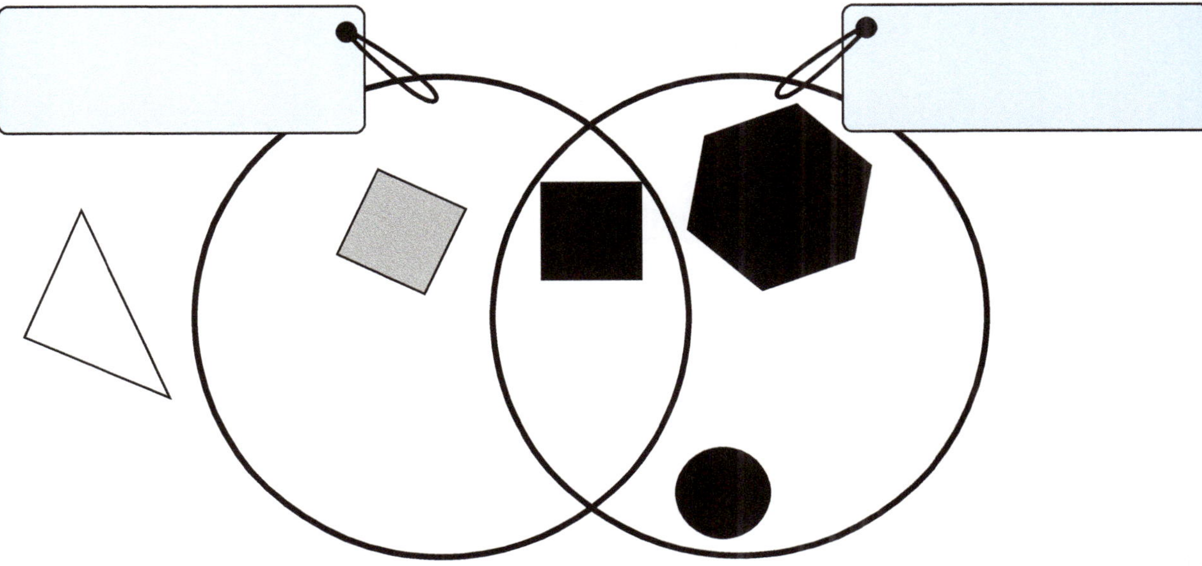

1. Label each loop.
2. Draw a new shape that belongs in each of the three sections inside the loops. (1 2 3)
3. Draw a new shape outside the loops.
4. The shape I drew in the middle belongs there because

____

____

____

Student Mathematician: ______________________ Date: __________

Angle-Sorting Tree

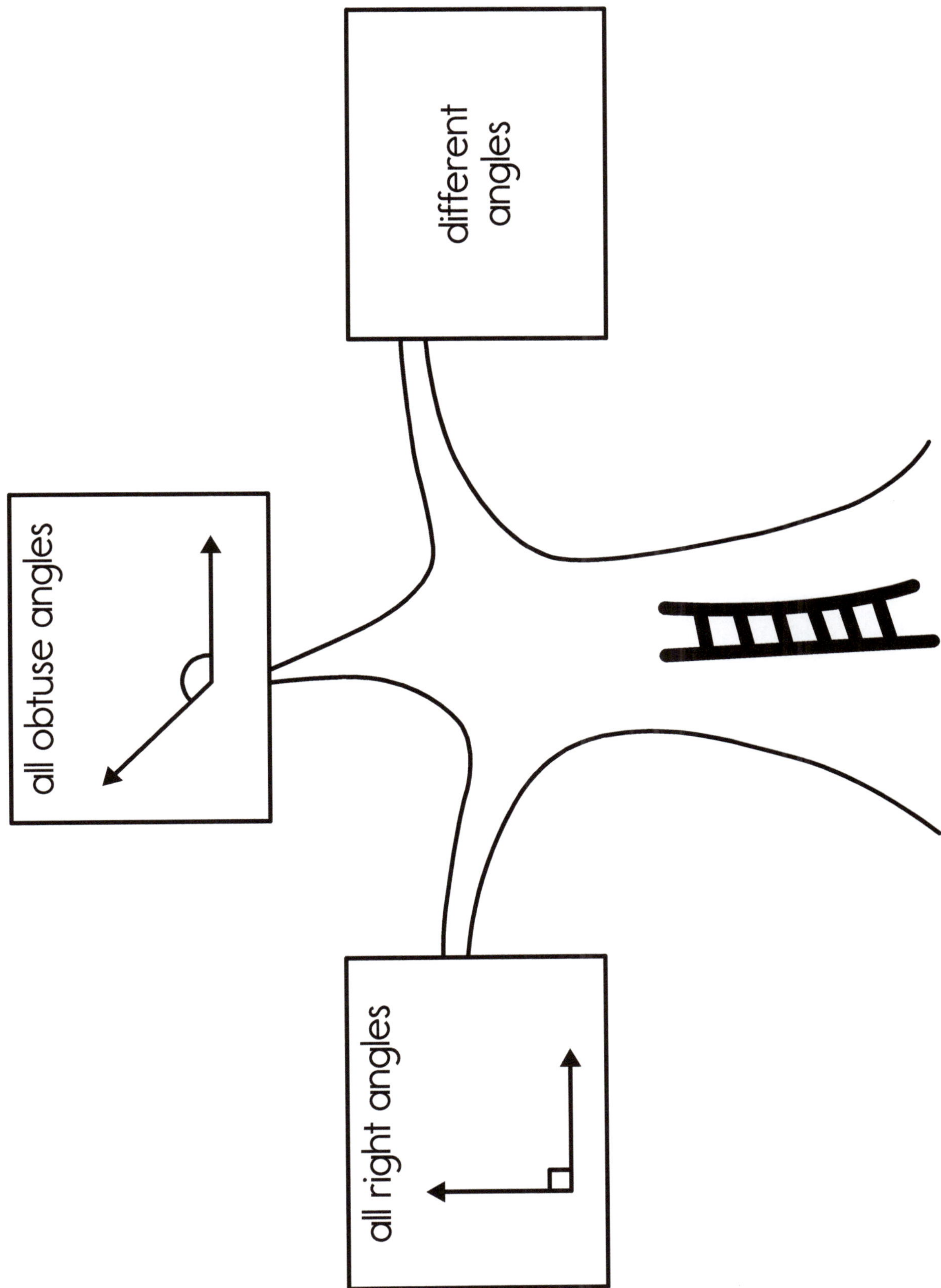

Student Mathematician: ______________________ Date: __________

Size- and Color-Sorting Tree

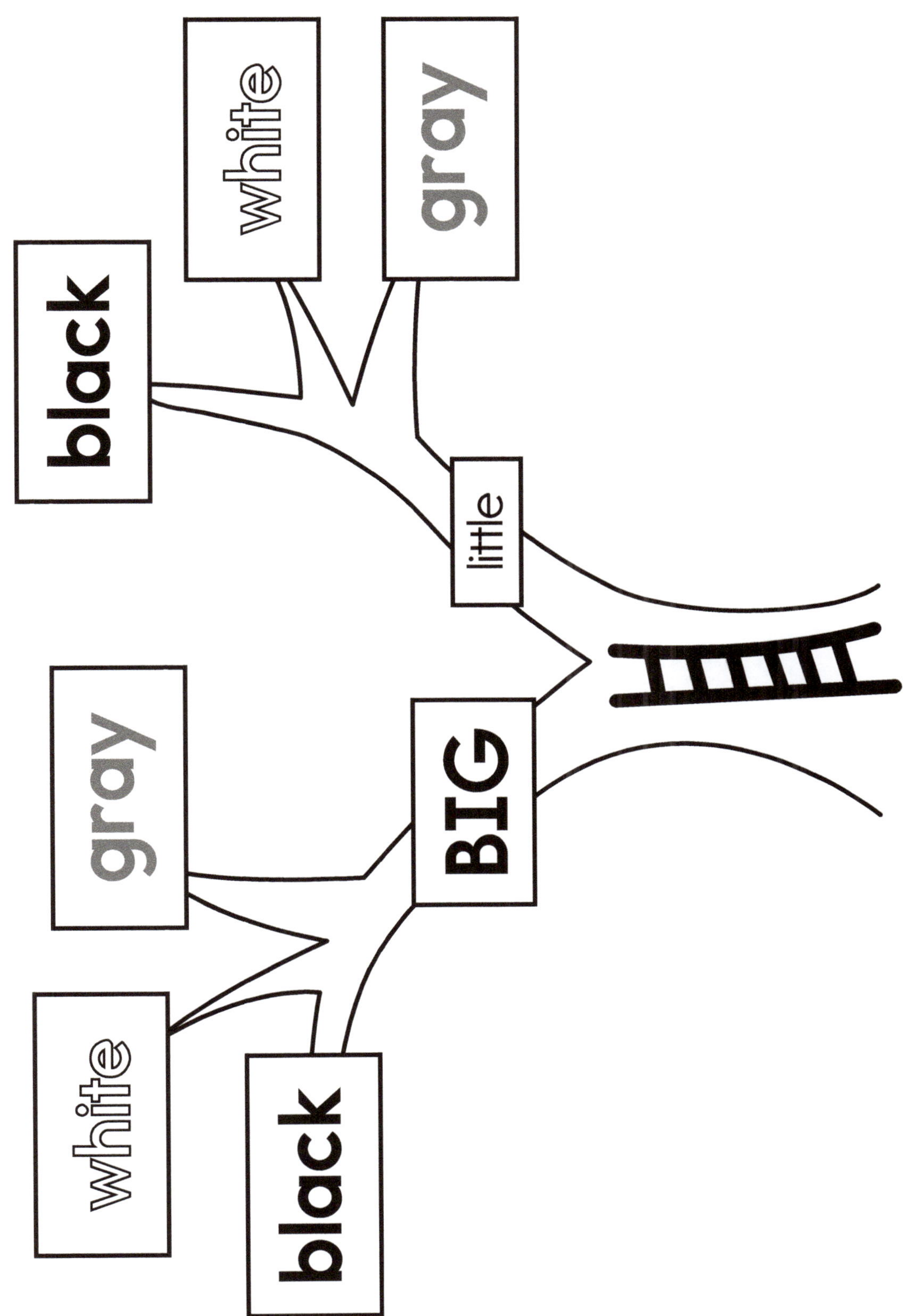

Student Mathematician: ______________________________ Date: ______________

Angle- and Size-Sorting Tree

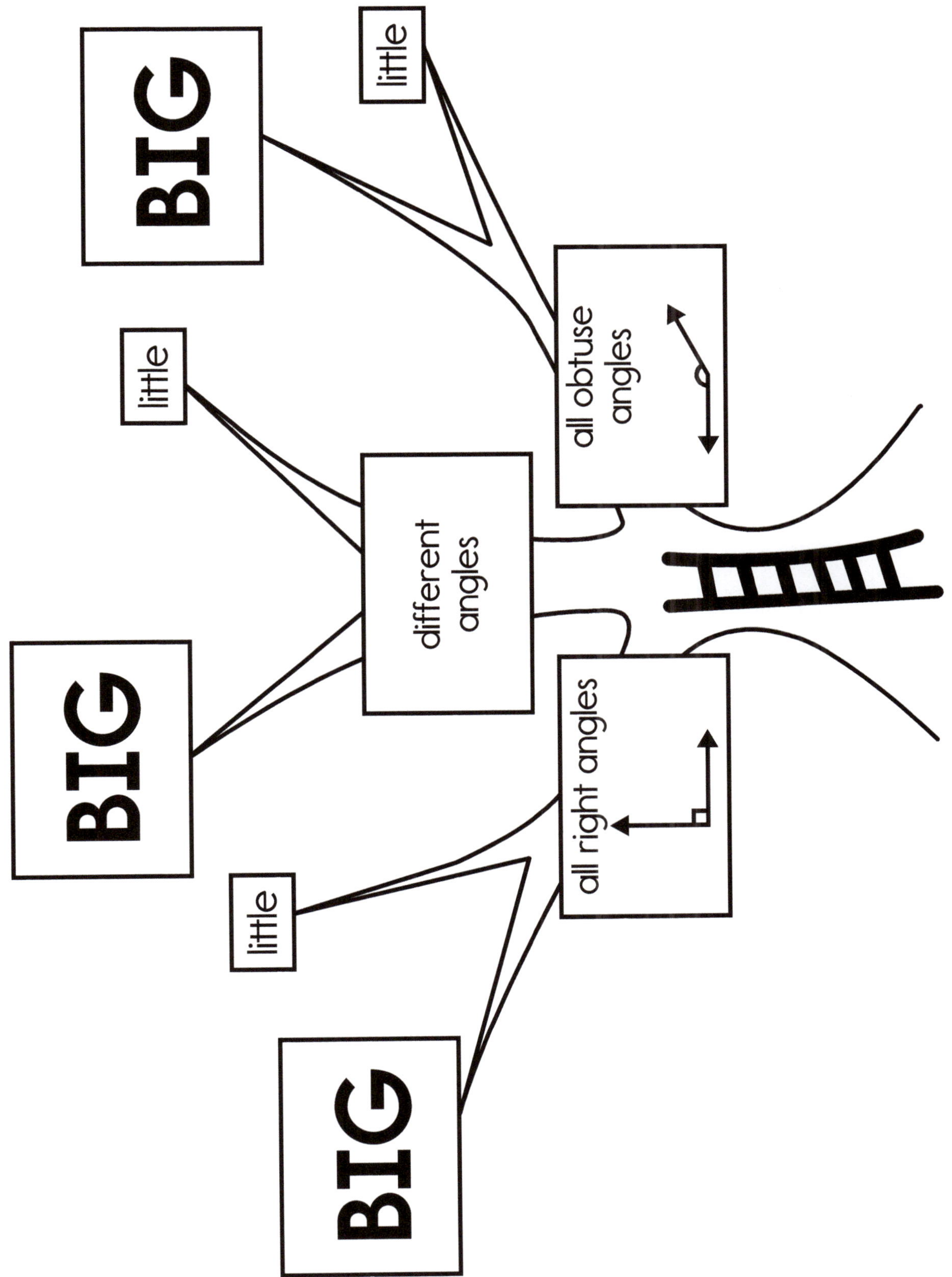

Student Mathematician: ______________________________ Date: ______________

What's Missing? — A

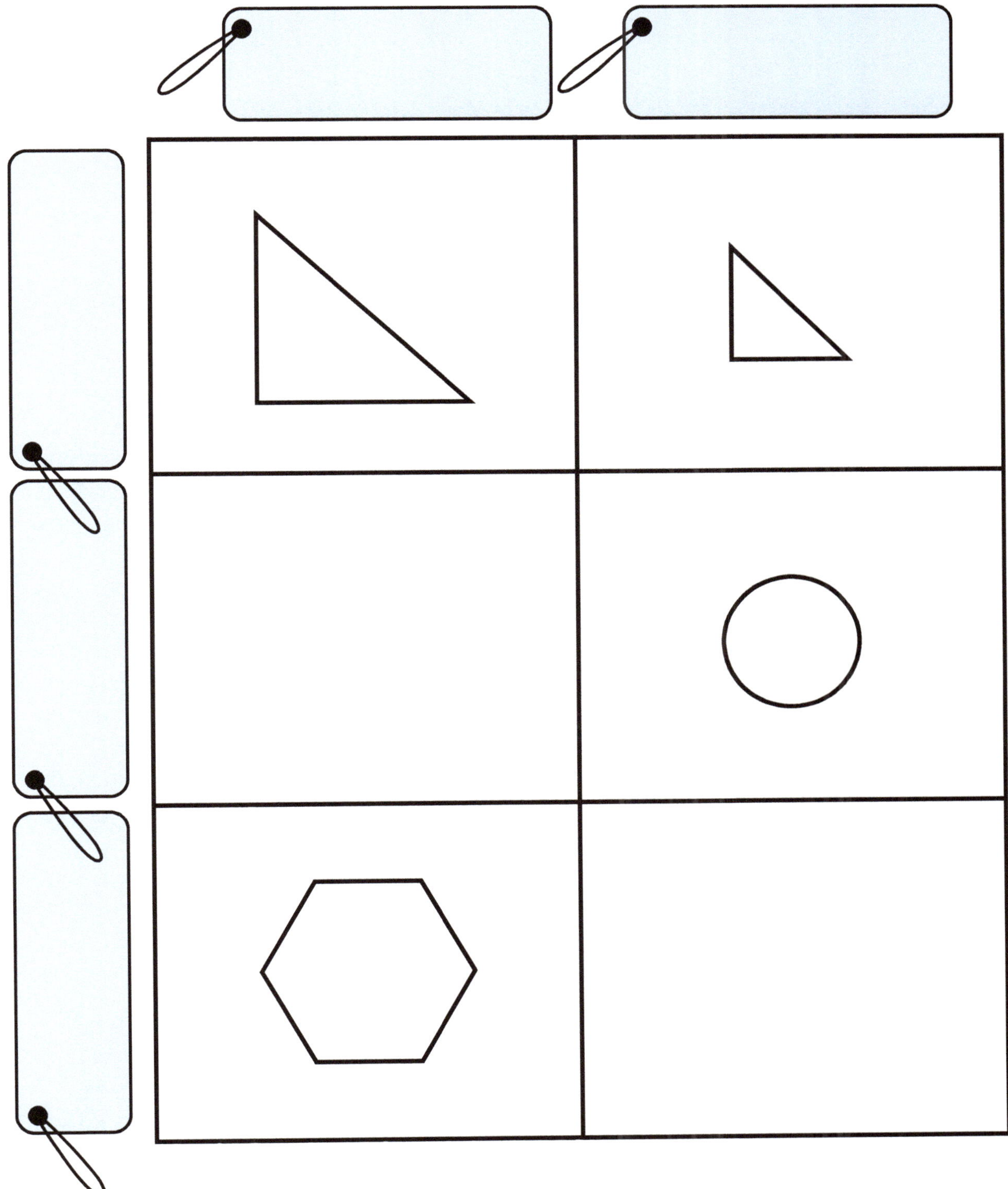

Student Mathematician: ______________________ Date: __________

What's Missing? — B

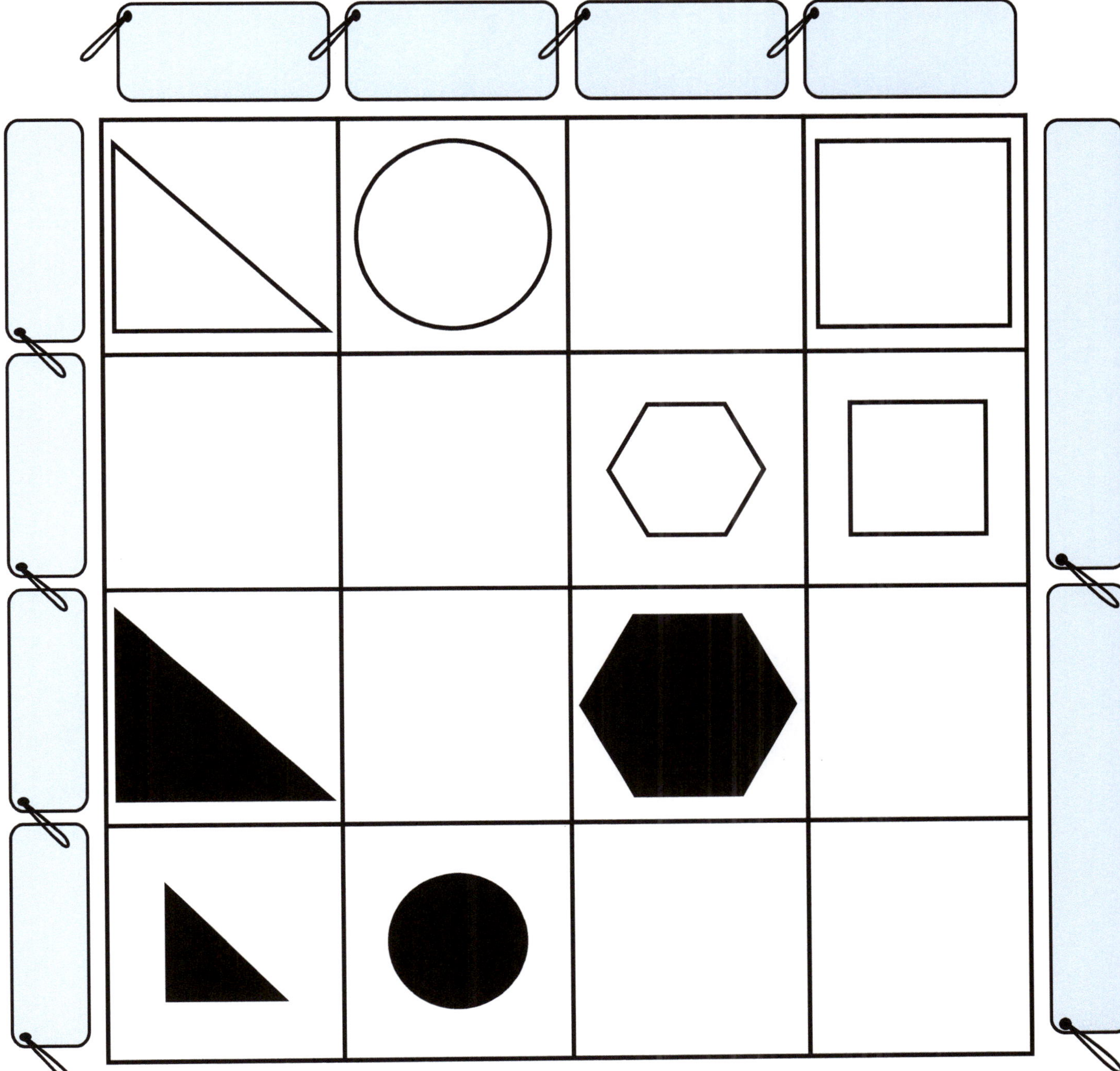

Student Mathematician: ______________________ Date: __________

What's Missing? – 1

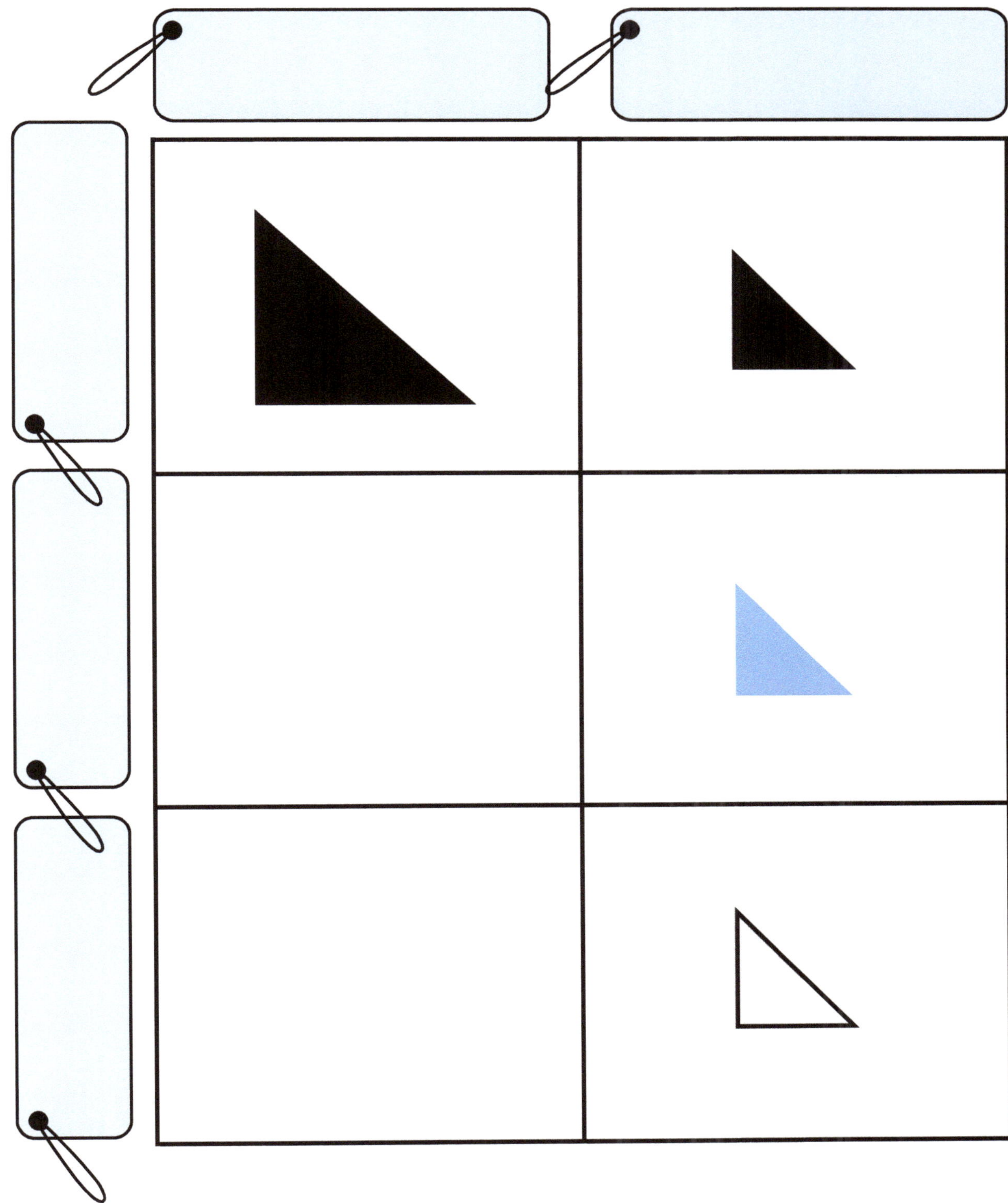

Student Mathematician: ______________________ Date: __________

What's Missing? — 2

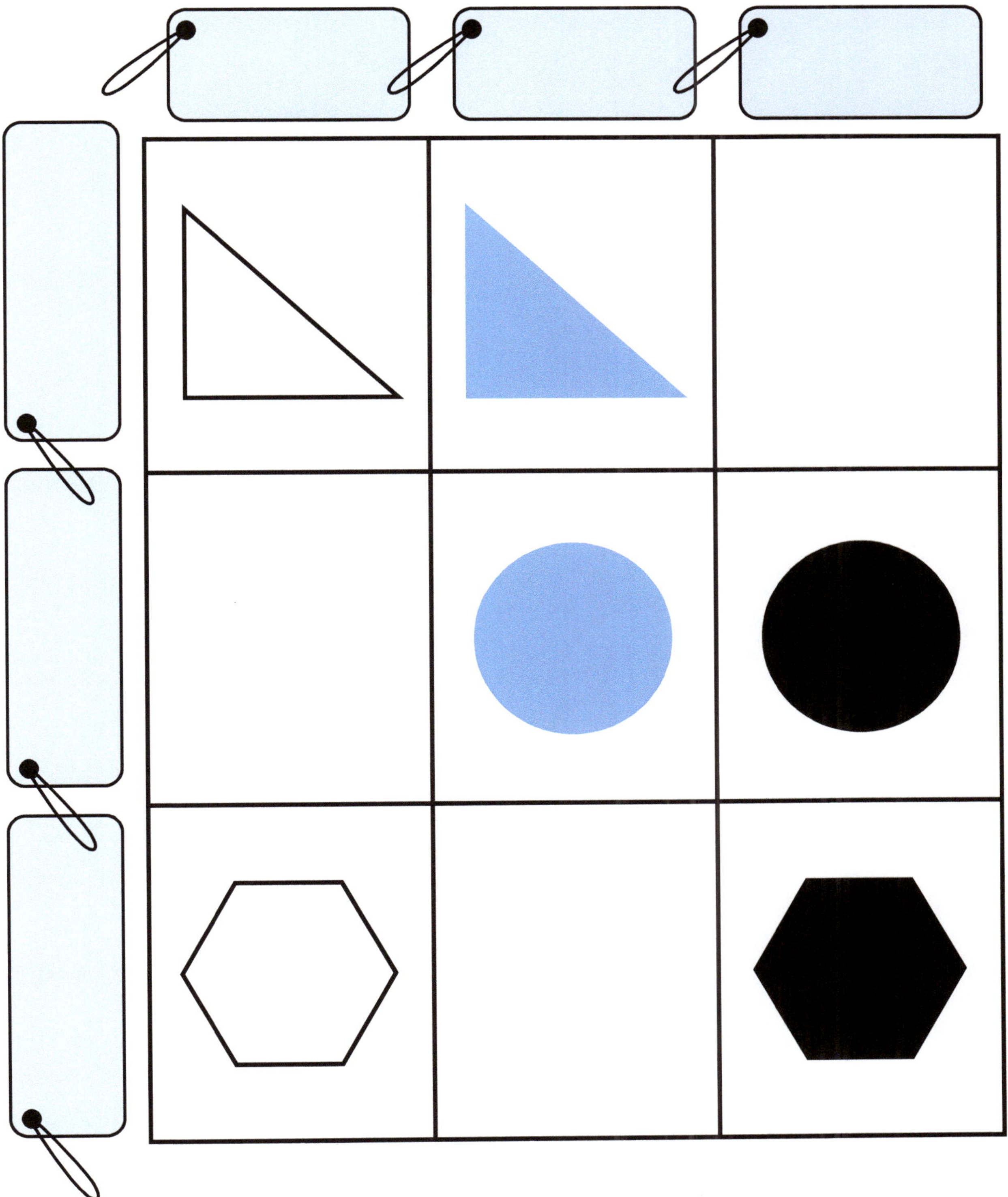

Student Mathematician: ______________________ Date: __________

What's Missing? — 3

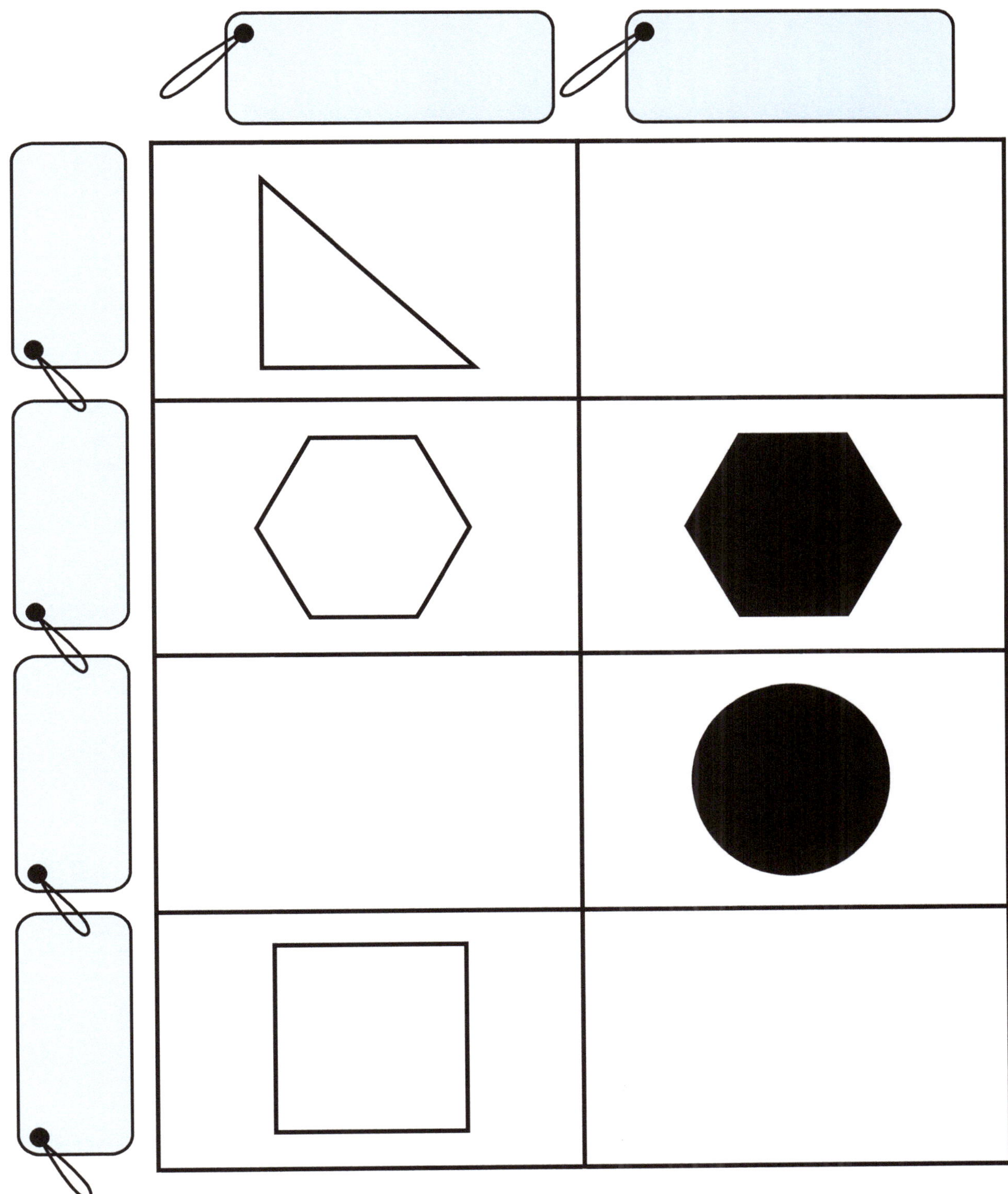

Student Mathematician: ______________________ Date: __________

What's Missing? — 4

Student Mathematician: ______________________ Date: __________

What's Missing? – 5

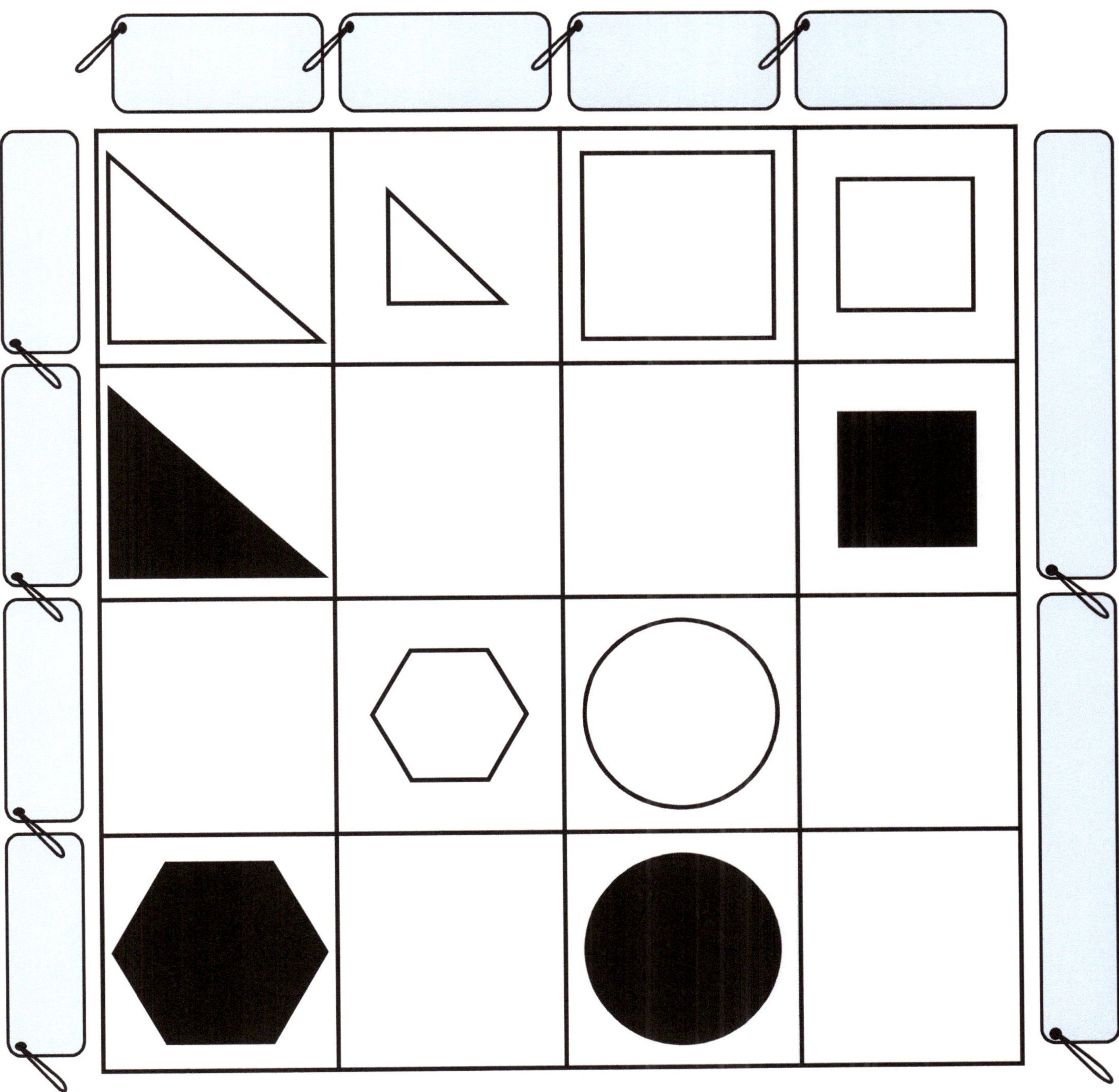

Student Mathematician: ______________________________ Date: ____________

What's Missing? — 6

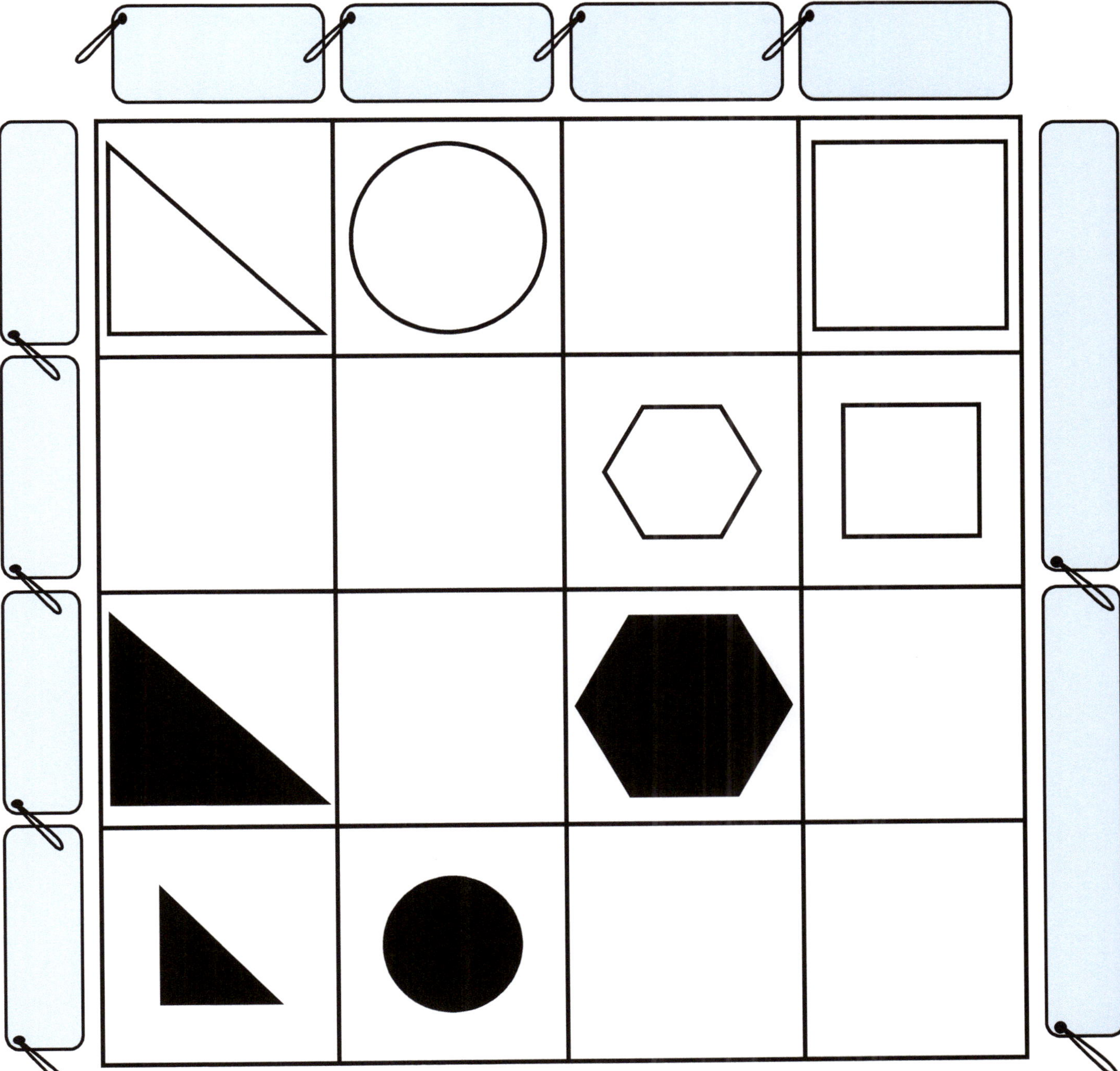

Student Mathematician: ______________________ Date: __________

Yes or No

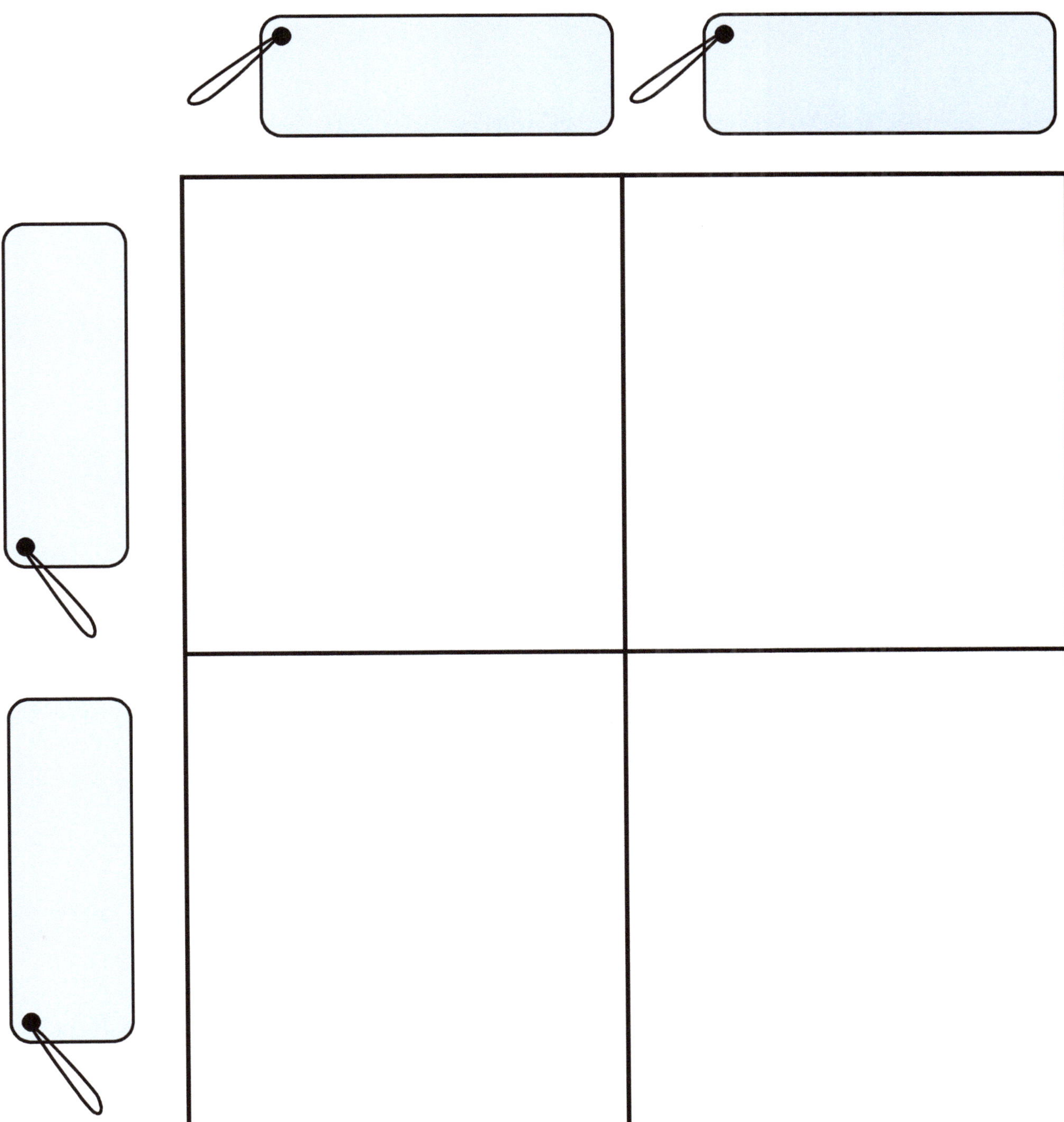

Imi and Zani
Amazon Birds Consulting

Student Mathematician: ______________________ Date: ____________

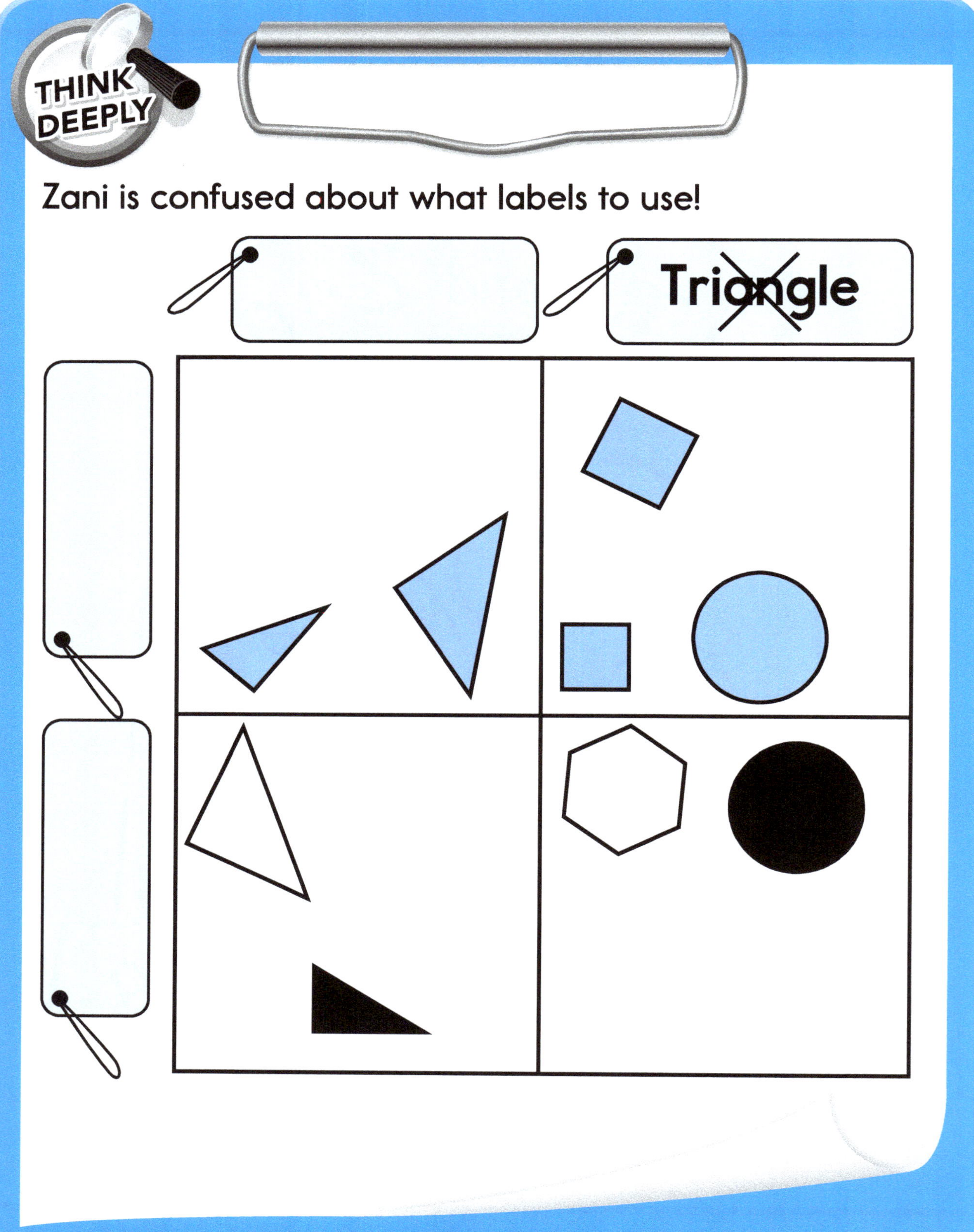

Student Mathematician: ______________________ Date: __________

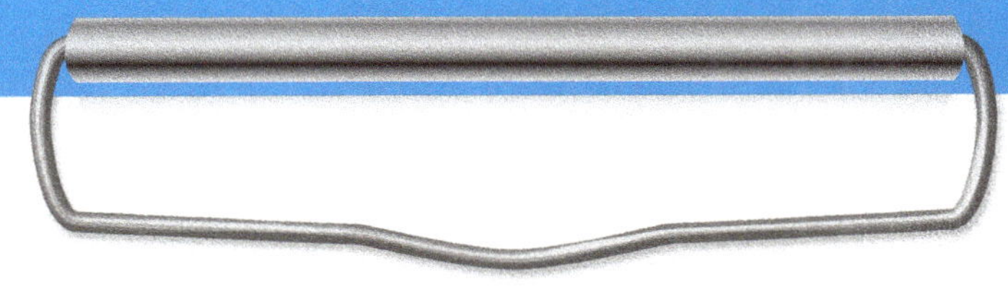

1. Label the rows and columns for me.

2. Draw one more shape in each box.

3. Look at the black triangle. Why does it belong in that box?

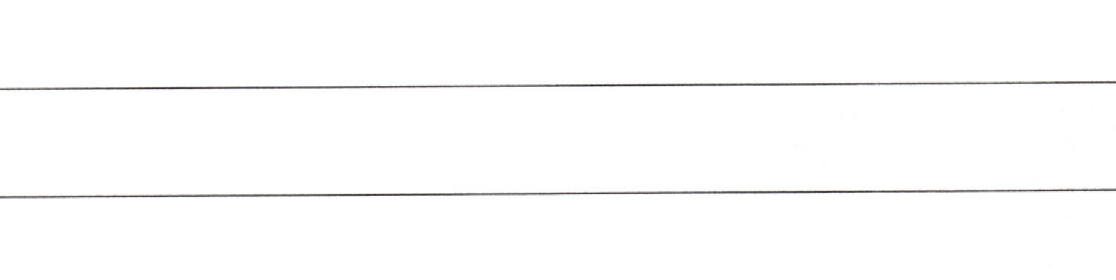

Student Mathematician: ______________________ Date: __________

Same Size and Shape?

Color each set of congruent shapes the same color.

Student Mathematician: ______________________________ Date: ______________

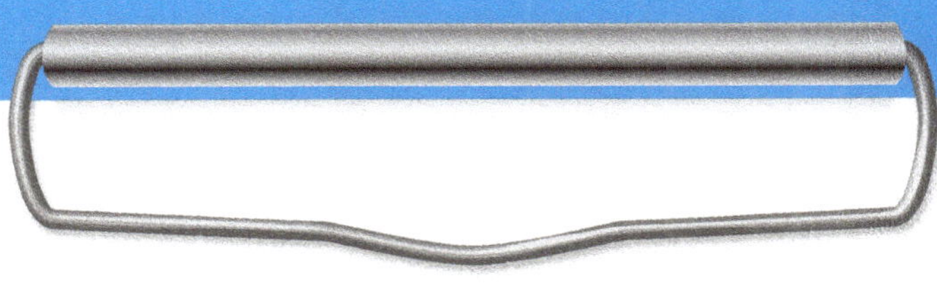

THINK DEEPLY

Help me, please!

1. Are these pictures congruent? Yes No

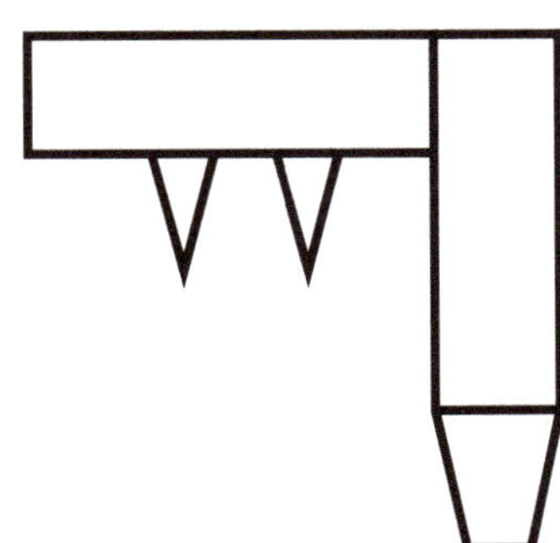

2. How do you know?

____

____

____

____

____

____

____

Student Mathematician: ______________________ Date: ____________

Where's Imi?

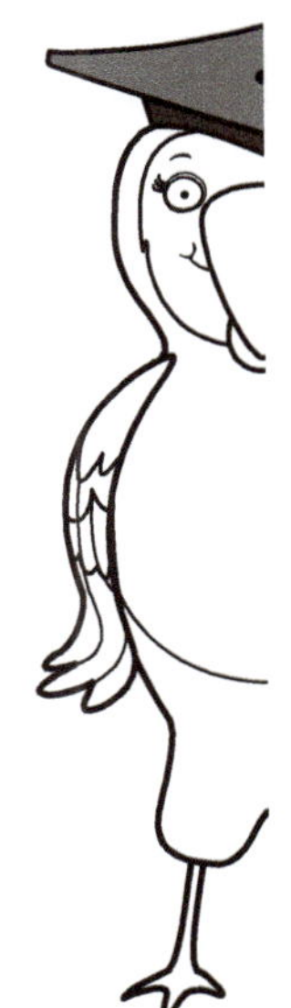

Student Mathematician: ______________________ Date: __________

Mirror Explorations

Make more bunnies. Make only turtles. Make silly figures!

Make the snake longer. Make it shorter. Make it disappear! Make another snake.

Make a taller building. Make a wider one. Make a shorter one.

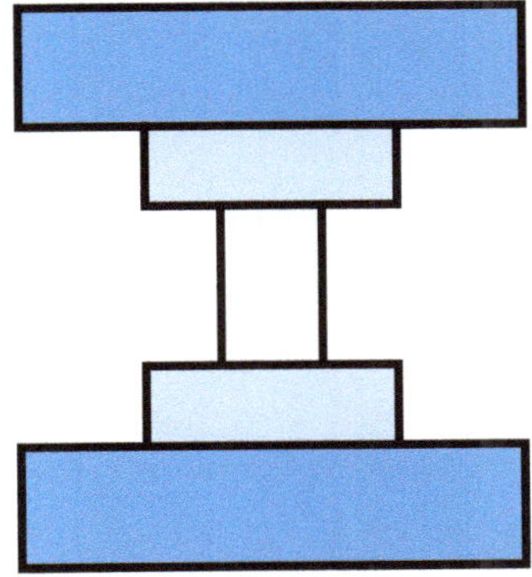

Student Mathematician: ______________________ Date: __________

Help the fish get out and add more grass.

Make a fish with many fins. Add more fish. Make a funny creature.

Make this number of dots: 1, 2, 3, 4, 5, 6, 7, 8, 9, then 10.

Student Mathematician: ______________________ Date: ____________

Do You See What You Started With?

Yes No

Yes No

Yes No

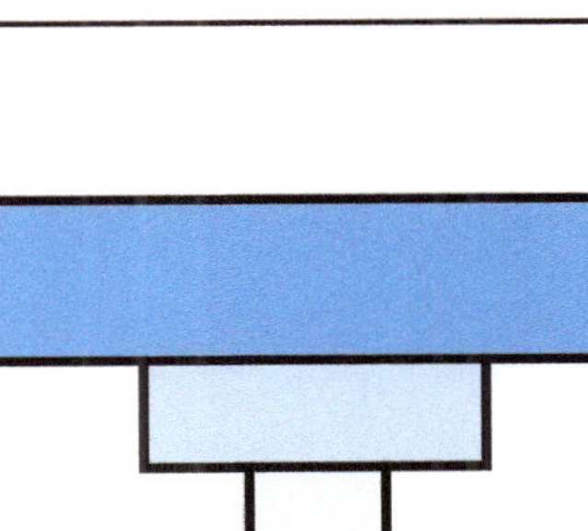

Yes No

Student Mathematician: ______________________ Date: __________

Symmetrical Shapes?

Student Mathematician: ______________________ Date: __________

Pattern Block Pictures

1. Make this butterfly. 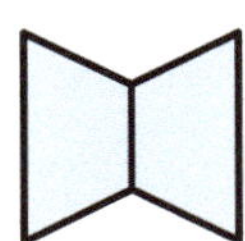

Do you see the same butterfly when you put the mirror on these lines?

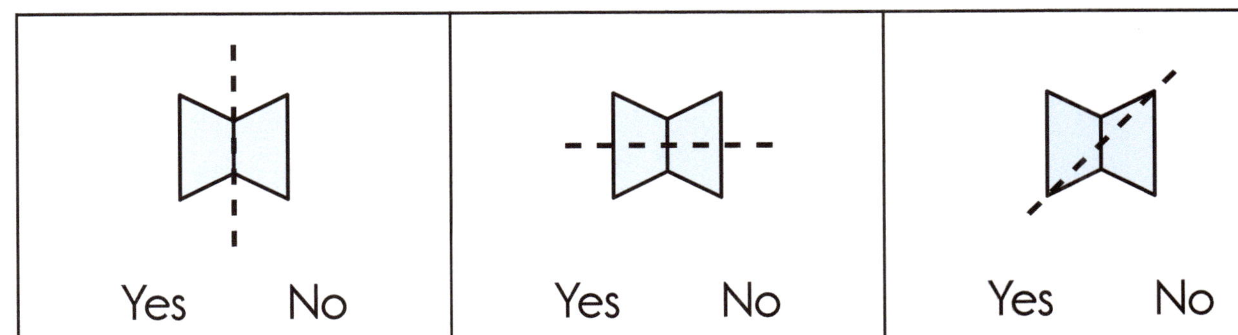

Is this butterfly symmetrical? Yes No

Circle how many lines of symmetry it has. 0 1 2

2. Make this flower. 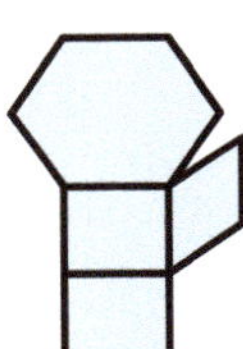

Do you see the same flower when you put the mirror on these lines?

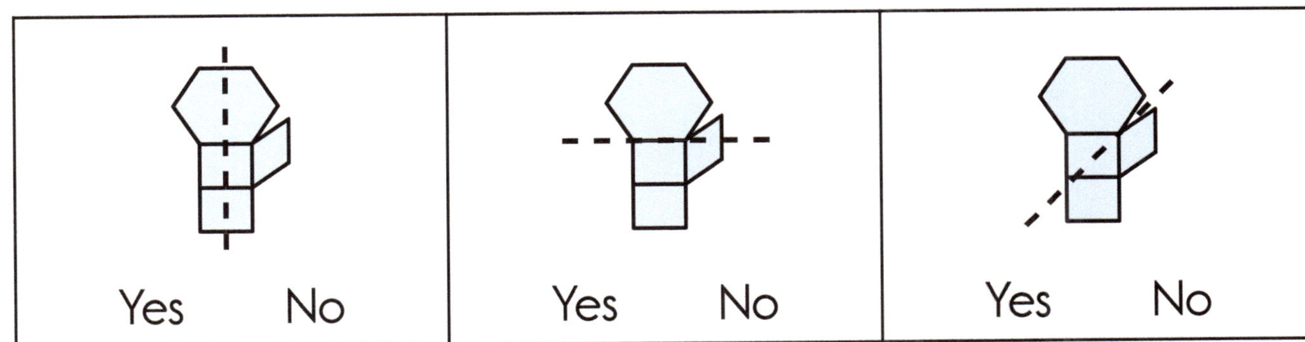

Is this flower symmetrical? Yes No

Circle how many lines of symmetry it has. 0 1 2

Student Mathematician: ______________________ Date: __________

3. Make this lamp. 

Do you see the same lamp when you put the mirror on these lines?

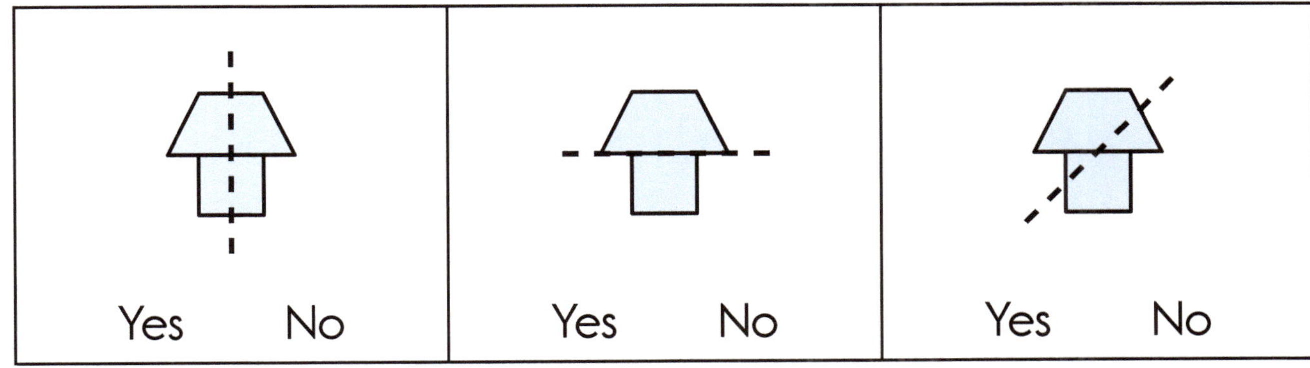

Is this lamp symmetrical? Yes No

Circle how many lines of symmetry it has. 0 1 2

4. Make your own design. Draw it here.

Do you see the same shape when you put the mirror on these lines?

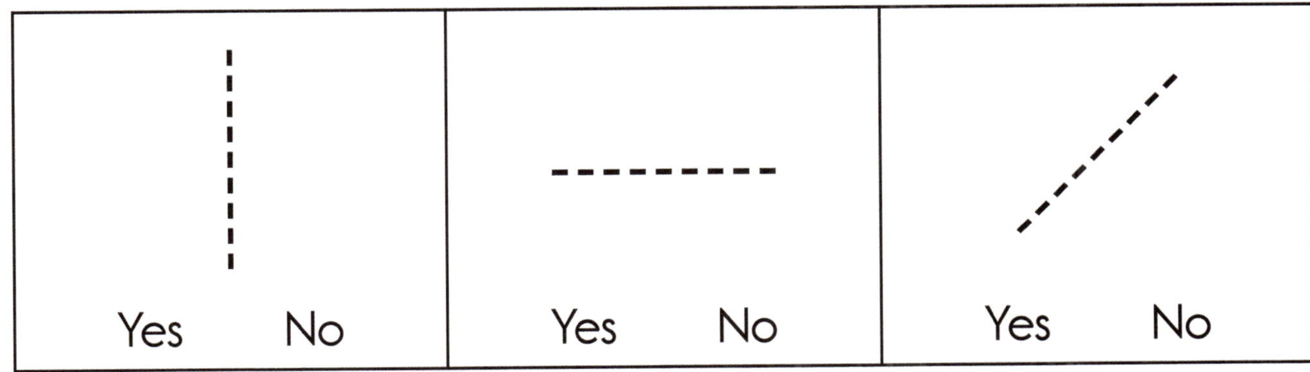

Is your design symmetrical? Yes No

Circle how many lines of symmetry it has. 0 1 2

Student Mathematician: ______________________ Date: __________

Pattern Block Symmetry Answer Sheet

Circle the shapes that show a line of symmetry.

A

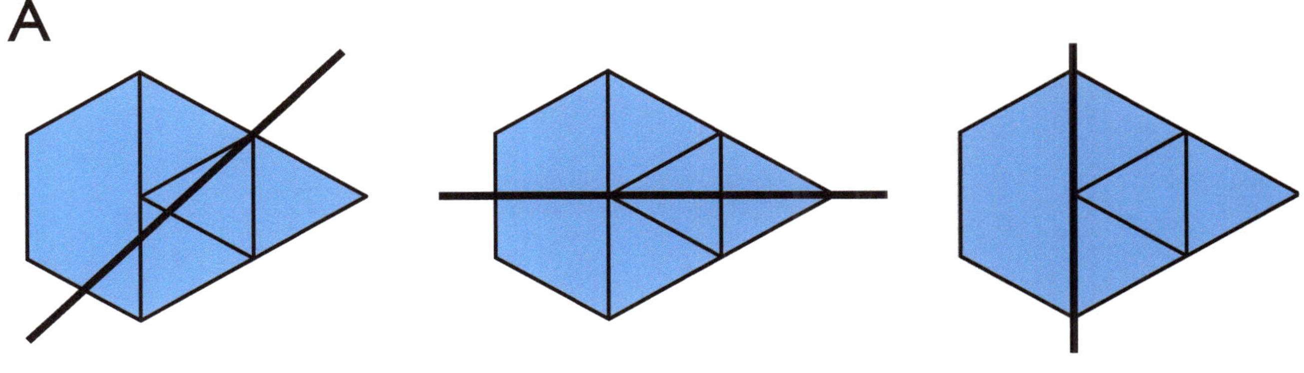

B

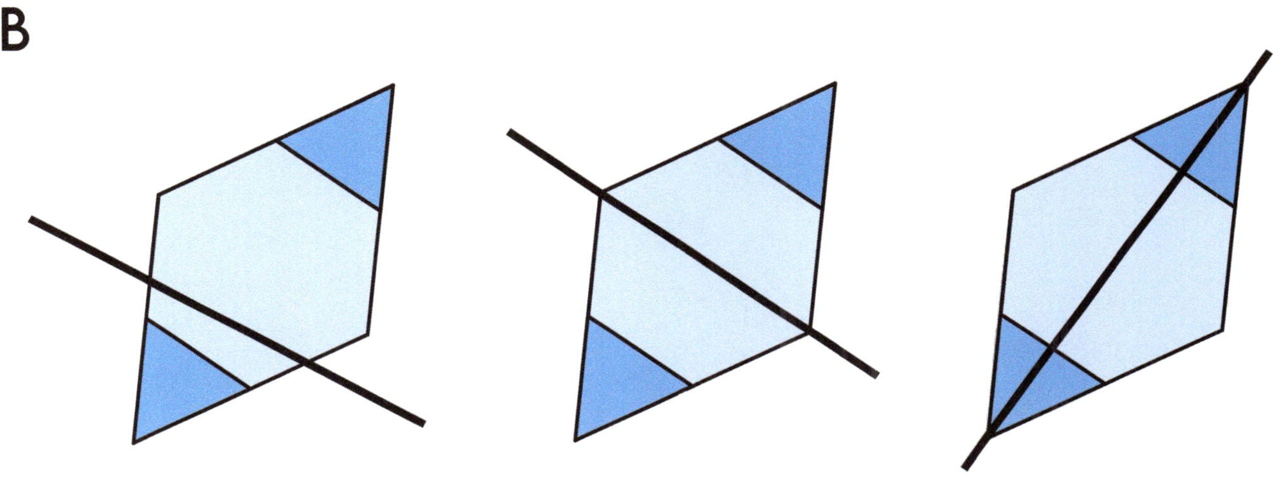

C

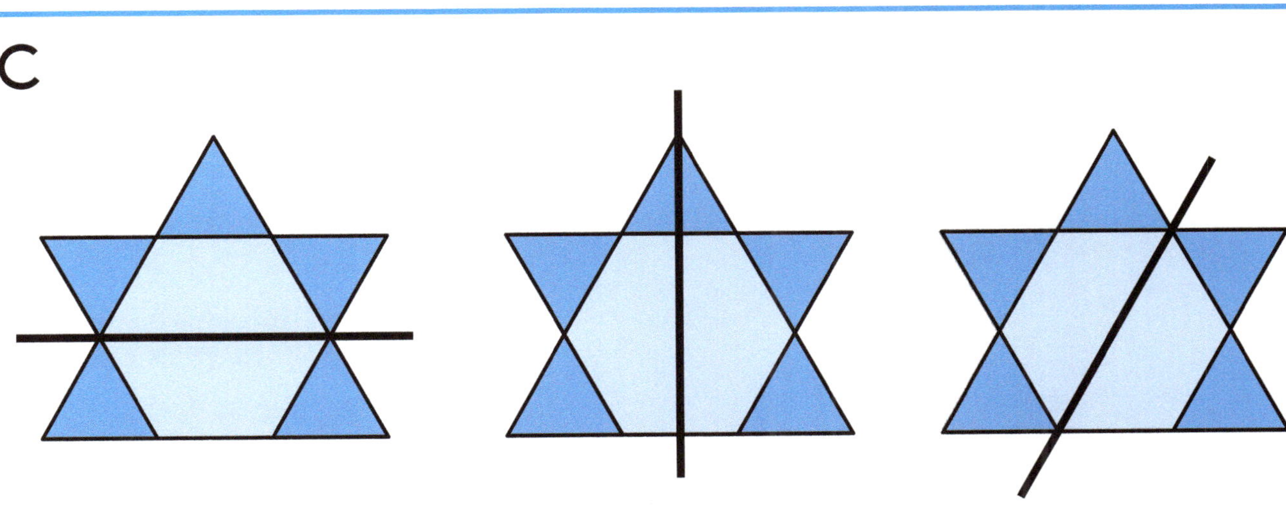

Student Mathematician: ______________________ Date: __________

D

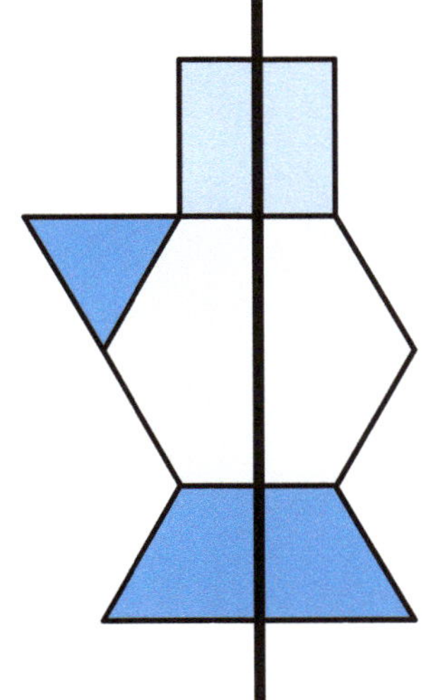

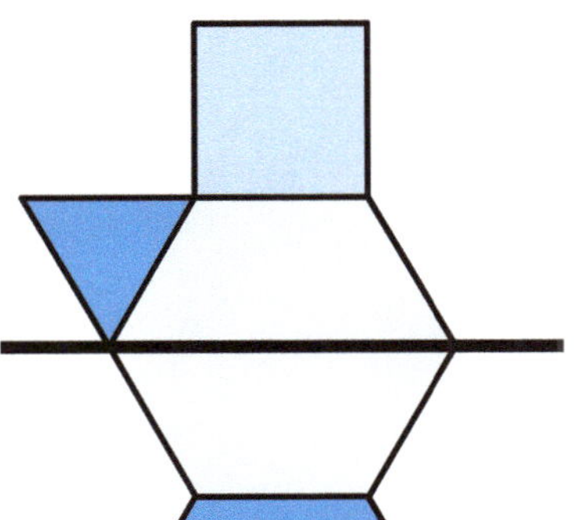

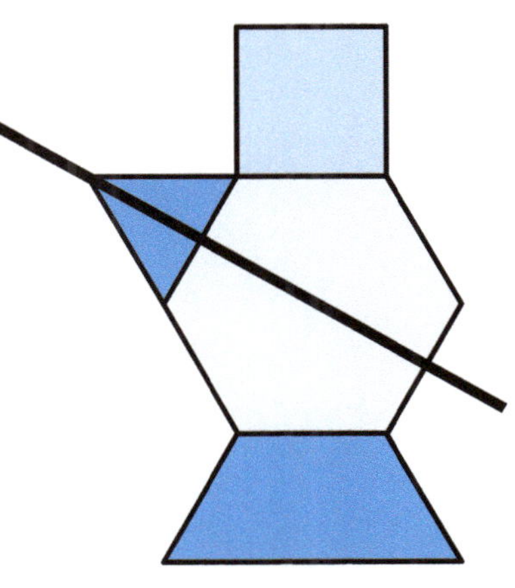

E

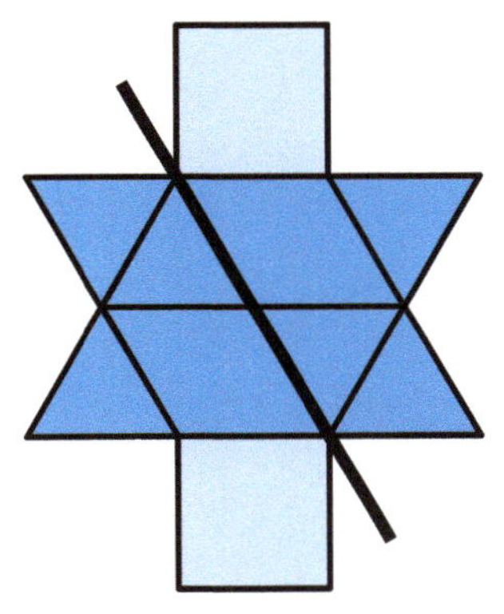

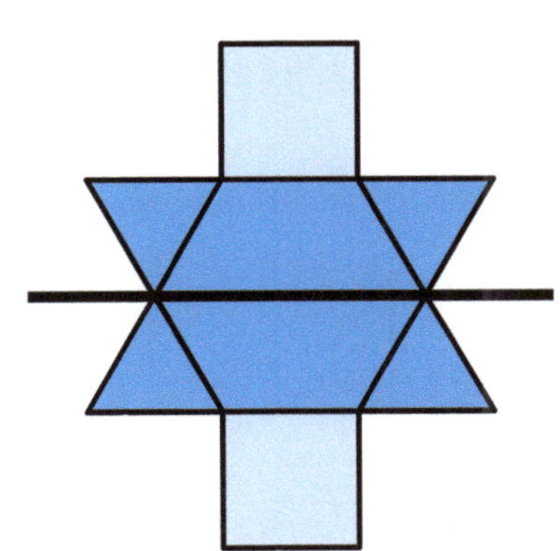

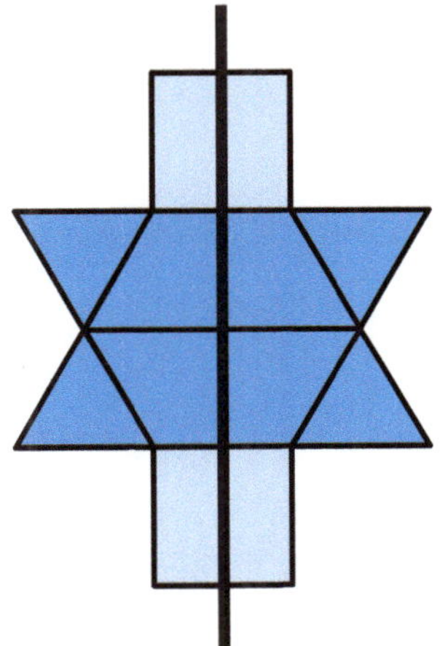

F

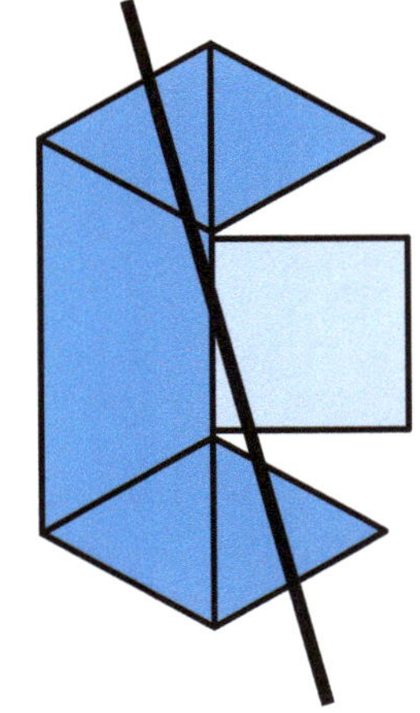

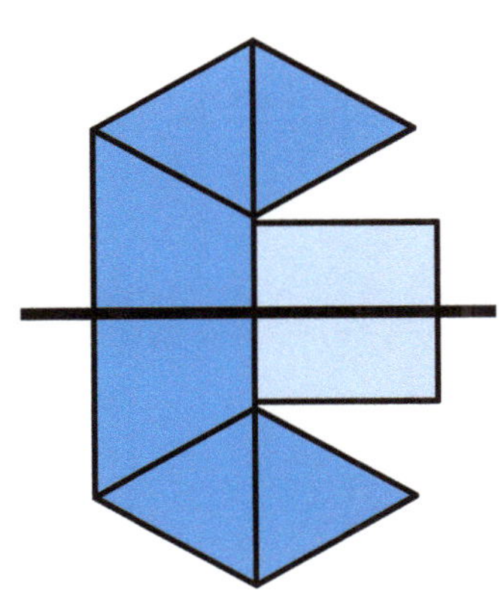

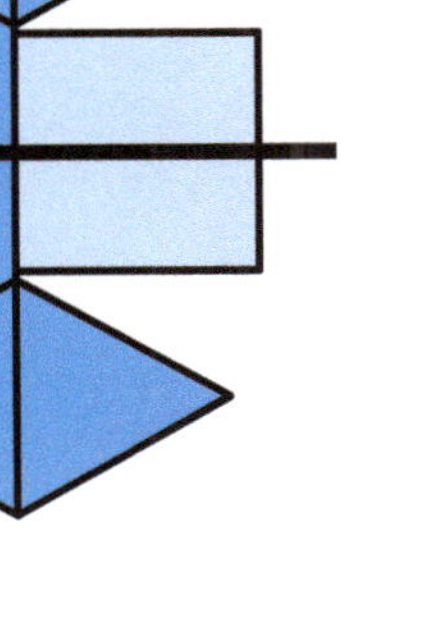

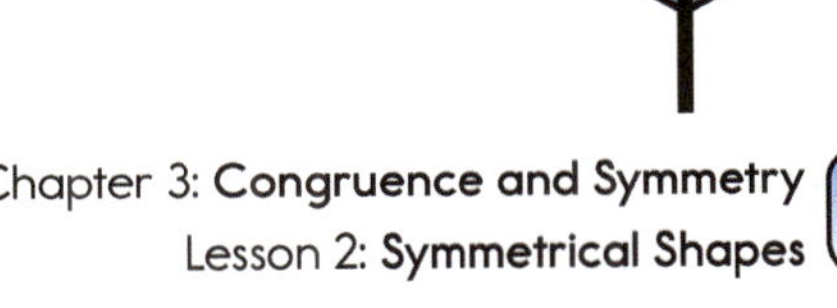

Student Mathematician: ______________________ Date: __________

THINK DEEPLY

Imi and Zani
Amazon Birds Consulting

I am confused!

Is this a line of symmetry?

Yes No

Draw a line of symmetry on the leaf. Explain why it is a line of symmetry. Tell me two different reasons.

Student Mathematician's Glossary

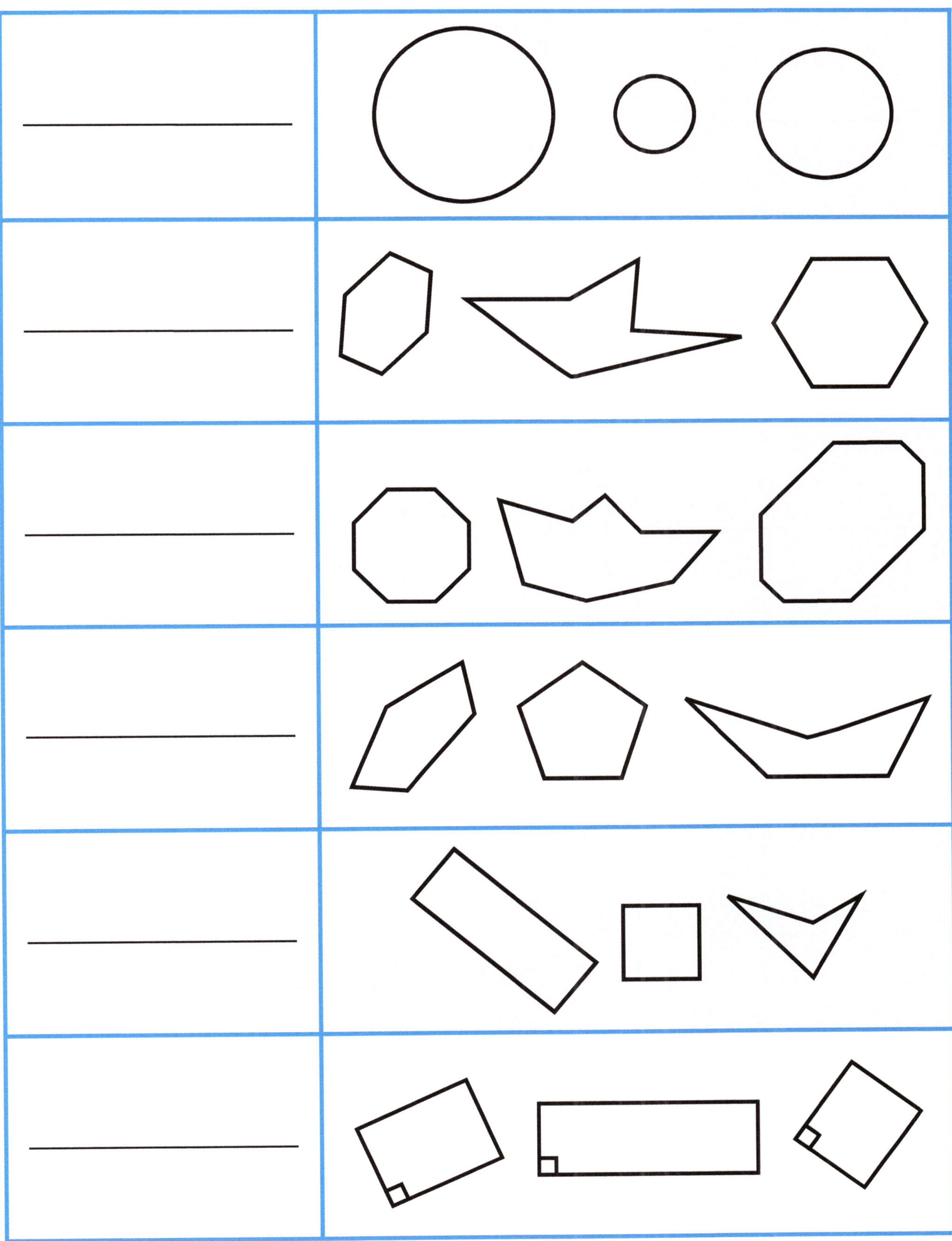

Student Mathematician's Glossary

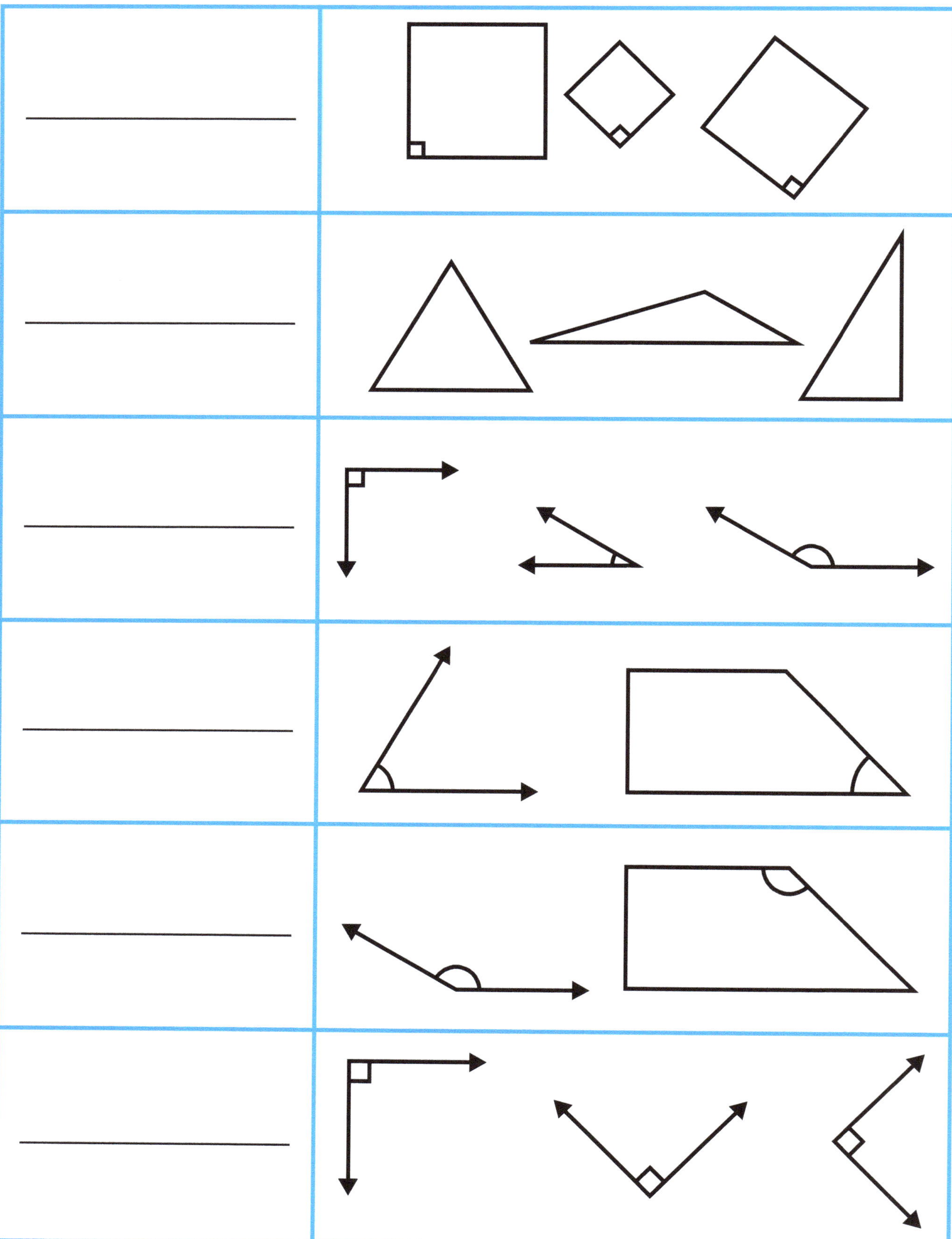

Student Mathematician's Glossary

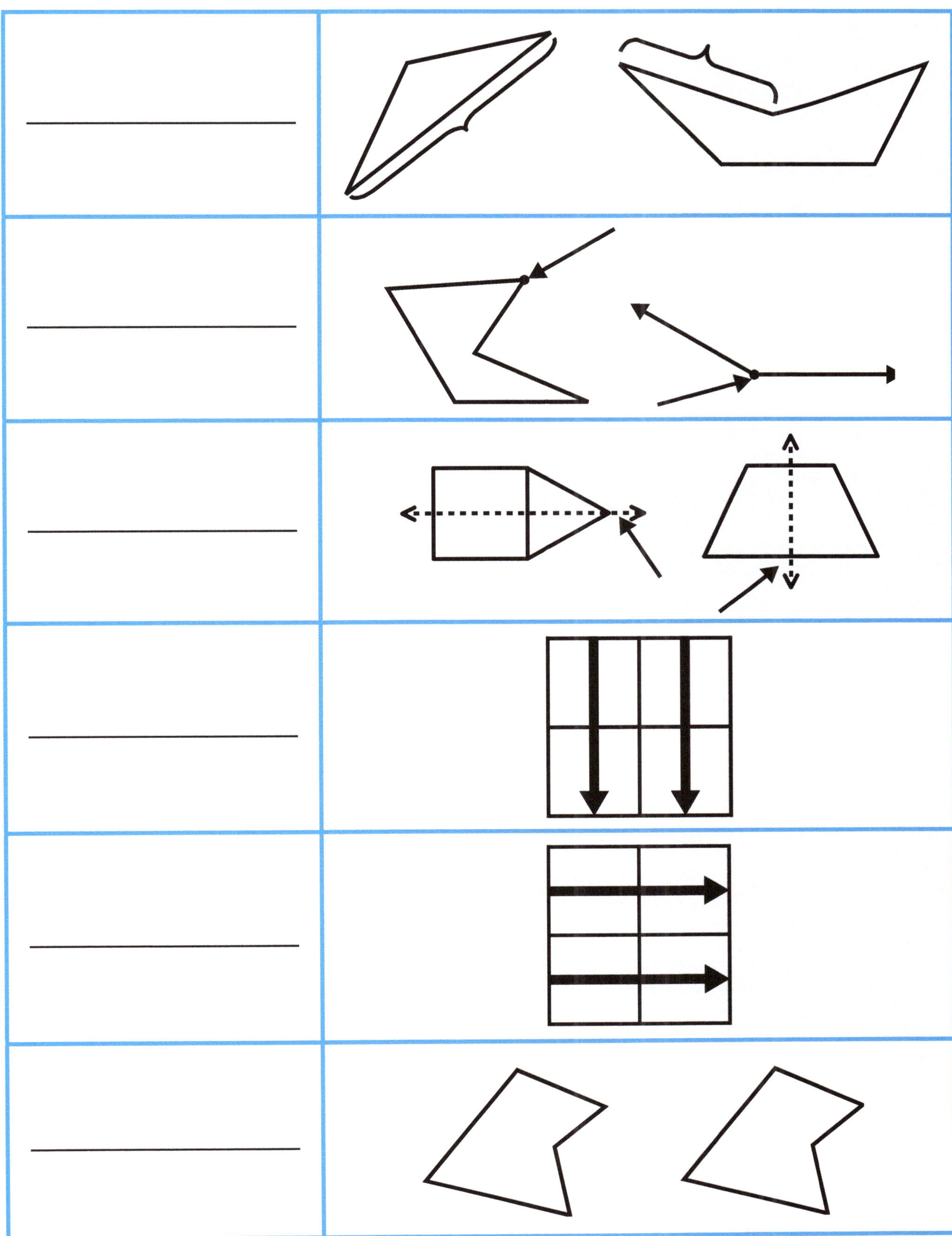

www.ingramcontent.com/pod-product-compliance
Ingram Content Group UK Ltd.
Pitfield, Milton Keynes, MK11 3LW, UK
UKHW050143280726
14058UKWH00006B/800

9 781524 931278